New under the Sun

New under the Sun

EARLY ZIONIST ENCOUNTERS WITH THE
CLIMATE IN PALESTINE

Netta Cohen

UNIVERSITY OF CALIFORNIA PRESS

University of California Press
Oakland, California

Library of Congress Cataloging-in-Publication Data

Names: Cohen, Netta, author.
Title: New under the sun : early Zionist encounters with the climate in
    Palestine / Netta Cohen.
Other titles: Early Zionist encounters with the climate in Palestine
Description: Oakland, California : University of California Press, [2024] |
    Includes bibliographical references and index.
Identifiers: LCCN 2023039659 (print) | LCCN 2023039660 (ebook) |
    ISBN 9780520397224 (cloth ; alk. paper) | ISBN 9780520397231
    (pbk. ; alk. paper) | ISBN 9780520397255 (ebook)
Subjects: LCSH: Jews—Colonization—Palestine. | Colonists—Palestine—
    Attitudes. | Public opinion—Palestine. | Immigrants—Palestine—
    History—20th century. | Palestine—Climate—Public opinion.
Classification: LCC DS119.7 .C634 2024 (print) | LCC DS119.7 (ebook) |
    DDC 320.54095694—dc23/eng/20231208
LC record available at https://lccn.loc.gov/2023039659
LC ebook record available at https://lccn.loc.gov/2023039660

Manufactured in the United States of America

33 32 31 30 29 28 27 26 25 24
10 9 8 7 6 5 4 3 2 1

To Jonah and his sister

Though we tend to think of knowledge as residing in minds, knowledge and ideas do not emerge from nowhere but from the interaction of human minds with specific places, materials, and things.

Linda Lorraine Nash, *Inescapable Ecologies : A History of Environment, Disease, and Knowledge*

Contents

Illustrations

Abbreviations

CEP	Committee for the Exploration of Palestine
CIAM	Congrès Internationaux d'Architecture Moderne
DCA	Department of Civil Aviation
GPA	German Palestine Association (Deutsches Palästina Verein)
JA	Jewish Agency
JCA	Jewish Colonisation Association
JNF	Jewish National Fund
MRU	Malaria Research Unit
PEF	British Palestine Exploration Fund
PICA	Palestine Jewish Colonization Association
ZO	Zionist Organization

Acknowledgments

A large part of this book has been written under the grey English sky. Getting geographical, climatic, and emotional distance from Israel/Palestine has allowed me to truly comprehend what it means to feel alien to a foreign climate. Yet, under the English sky I have also made new discoveries and experienced unexpected intellectual developments, and these could not have become fruitful without the institutions and people who supported, stimulated, and enriched my work (here and elsewhere) during the last years.

First and foremost, I am grateful for the guidance, encouragement, and support of my former PhD supervisor, Derek Penslar, who believed in my sometimes-messy ideas from the very beginning and who has helped in developing these ideas into a coherent written work. In addition, I wish to express my deepest appreciations to the generous Mark Harrison, Erica Charters, Rob Iliffe, Marie-Aline Thebaud-Sorger, Jonathan Schorsch, Eitan Bar-Yosef, Yaacov Yadgar, Yfaat Weiss, Daniel Wildmann, Elisabeth Gallas, Gadi Algazi, Areej Sabagh, Yael Zerubavel, Haim Yacobi, Dafna Hirsch, David Schorr, and Noah Hysler-Rubin for kindly agreeing to read and comment on parts of my work and for being open to engaging in insightful and enjoyable conversations on climate, medicine, science, technology, Jewish, Zionist, and Israeli-Palestinian history.

I owe a particular debt to the generosity of Christ Church College for supporting my research in the last four years. Other institutions that funded and braced my research on this project were the Pears Foundation, St. Antony's College and the Middle East Centre, the Center for Jewish History in New York City, the Leo Baeck Institute and the Studienstiftung des deutschen Volkes, and the Taub Center for Israel Studies at New York University. I gained tremendous experience and knowledge from working with the Jewish Environmental History research group at the Leo Baeck Institute and from participating in the Colonial Landscapes research group of the Minerva Institute for German History at Tel Aviv University.

During my work on this project, I was fortunate to receive the professional and personal aid and support of many friends and colleagues. Thank you, Harriet Mercer, Efrat Gilad, Daniel Herskowitz, Dominik Huenniger, Tamar Novick, Nimrod Ben Zeev, Basma Fahoum, Aviv Derri, Atalia Nevo-Israeli, Natalia Gutkowski, Eitan Bloom, Kate Lebow, Paul Betts, Sarah Dry, Michael Willis, Liat Kozma, Shira Wilkof, Shira Pinhas, Moti Golani, Lina Baruch, Dotan Greenvald, Dotal Halevi, Tami Tsarfati, Yoni Mendel, Ray Schrire, Ella Elbaz-Nir, Amit Levi, Roii Bell, Vincent Di-Piazza, Louise Chapman, Noa Shauer, Linoy Elhanani, Tal Refaeli, Ofer Salman-Sagie, Smadar Yaniv, Michal Kra, Nir Shalev, Jaclyn Granick, and Maayan Ravid. Lastly, I wish to thank the Huguet and the Cohen families, and especially Michael, Ima, and Abba for believing in me and supporting me endlessly. My utmost gratitude goes to the pillars of my everyday life who force me to stay alert and engaged, who make me laugh (a lot!), and who remind me of what truly matters—Benjamin, Jonah (and his sister), to whom this book is dedicated with love.

Introduction

THE COLONIAL HISTORY OF CLIMATE
INVESTIGATION IN PALESTINE

The aim of this book is to examine networks of human and non-human connections pertaining to the climate in Palestine in the first half of the twentieth century. To borrow Bruno Latour's words, these networks will be viewed as *"simultaneously real, like nature, narrated, like discourse, and collective, like society."*[1] More specifically, this book tells the story of climate investigation in Palestine and the way it shaped and transformed the life of humans and non-humans in this country. Likewise, it examines how climate inspired experts' response, and why this response was not always successful. It is in part a story of people and their old and new environments, as well as a story of power relations and ecological engagements. Those who dwelled in Palestine in the first half of the twentieth century—Palestinians and Jews—had different experiences of and engagements with their local climate and environment. While these experiences were inseparable, they also reflected very different perspectives that were a result of their distinct colonial positions.

This study focuses mainly on the attempts of Jewish European[2] experts to understand, manage, and deal with the climate in Palestine while highlighting the link between their orientalist views towards both the local climate and environment as well as towards its Palestinian population.

Orientalism, as Edward Said defined this term, "expresses and represents the Orient culturally and even ideologically as a mode of discourse with supporting institutions, vocabulary, scholarship, imagery, doctrines, [and] colonial bureaucracies."[3] According to Said, orientalist knowledge usually also tends to transform into a practical plan for domination.[4] More recently, within the field of environmental history, Diana Davis has defined environmental orientalism as a Western representation of Middle Eastern and North African environments that were viewed as "strange and defective" compared to Europe's "normal and productive" environment. This view resulted in the perceived need to "normalize" and "repair" them, thus justifying imperial projects of agriculture, irrigation, and forestation.[5] Davis's work, published in 2011, was one of the first to interrogate the power relations and forms of knowledge production concerning nature in this region from a postcolonial perspective.[6] While Davis used the concept of environmental orientalism to refer mainly to British and French colonial environmental approaches, the current study wishes to highlight Zionist orientalist interactions with the climate and environment in Palestine and its discourses about it. By "discourse" I do not only include how climate was discussed in words, but also how it was formulated within social, political, cultural, scientific, and very much material spheres.[7]

Addressing climate via an environmental orientalist approach goes hand in hand with adopting a more general colonial analytical framework, which is essential to this study for several reasons. The most obvious reason is related to the fact that climate science, and especially the investigation of warm climates by Western societies in the modern period, has been deeply embedded in a colonial context. Historians have often addressed European colonial perceptions of warm climates and suggested that the development of climate as a field of intense medical and scientific study was largely a result of European encounters with climates outside the so-called "temperate zone."[8]

As these historians demonstrate, the scientific study of warm climates by Europeans had already received growing attention with the first colonial conquests of the "New World" in the late fifteenth century. Nevertheless, climate obtained even greater attention from the eighteenth century onwards as a result of a number of scientific, economic, and political developments. It was predominantly the increasing transportation of humans, as well as of animals and plants, between the metropoles and their colonial holdings, and the enormous financial possibilities enabled by global and imperial commerce, that led European trading companies during this period to hire experts to study environmental and climatic conditions necessary for human, animal, and plant acclimatization.[9]

However, as Deborah Coen and Martin Mahony individually show, the successful quantification of climate required numbers in different times and places to have the same meaning. Thus, it was only during the era of high imperialism in the second half of the nineteenth century—with its advanced communication and transportation technologies—when information on climate achieved its zenith, becoming standardized, widespread, and prevalent.[10] The German-Jewish meteorologist Rudolf Feige confirmed the relationship between colonialism and climate science when he wrote that "every colonial enterprise, wherever it may be, relies tremendously on the climatic potential of the target country, and all colonialists must therefore notice the recordings of weather observations."[11]

Today most scholars agree on the colonial characteristics of the Zionist project.[12] Moreover, in recent years new scholarship has started to tackle the environmental aspects of this colonial project.[13] In 2023 Irus Braverman coined the term "settler ecologies" to describe "the structural as well as the plural and dynamic components in of the colonial administration of nature as configured through scientific modes of knowledge and practice" in Israel/Palestine.[14] One of the main objectives of this study is, therefore, to enrich and deepen this understanding by specifically addressing the role of climate within early Zionist settlement schemes and strategies. Zionist approaches to the climate in Palestine during the first half of the twentieth century were certainly formed by colonial science. Nevertheless, as we shall see shortly, the Zionist employment of colonial knowledge concerning climate was not one-dimensional, and it often revealed a blend of diverse colonial and metropolitan notions and tactics concerning the local environment and its Arab population.

For example, during the early twentieth century the Zionist Organization (ZO) regularly hired the services of "foreign" experts for its scientific expeditions (mainly British and German scientists, but also several Belgian, Swiss and French experts) who tended to employ Western research methods and axioms to examine climatic conditions that were new and unfamiliar to them. Instead of trying to access local knowledge, these experts usually preferred using standardized scientific instruments such as thermometers, barometers, and rain gauges. They often stored this data in tables and charts and then used statistical tools to remotely analyze their findings.

This type of knowledge concerning warm climates was also mirrored in the education and work experience of Jewish Zionist experts themselves. At the turn of the twentieth century, Jewish experts generally acquired their academic studies in European universities, where they were often exposed to knowledge concerning warm environments that had been developed in the colonies. In addition, some of these experts had even participated in British and German colonial enterprises before joining the Zionist movement, and in some cases, they retained other colonial affiliations even after becoming Zionists.

As Derek Penslar and other historians have already shown, the German-Jewish botanist and Zionist leader Otto Warburg can serve as a prime example of this trend. Born in Hamburg in 1859, Warburg studied botany at the Universities of Bonn, Berlin, and Strasbourg. Following his graduation, he embarked on a three-year scientific excursion to study tropical plants in Asia and Australia. In 1897, he helped to found the German Colonial Economic Committee (*Kolonialwirtschaftskommitee*), which focused on promoting economic development in the German colonies, and became one of its most active members. Soon afterwards he founded the Committee's scholarly journal on colonial agriculture, entitled *The Tropic Planter* (*Der Tropenflanzer*), which he also edited.[15]

Warburg's prominent position in the German colonial scientific enterprise did not, however, prevent him from being a zealous Zionist. Warburg joined the Zionist movement in 1898, and from 1903 onwards he held several important executive positions in its organization, including as director of the Committee for the Exploration of Palestine (CEP) and chair of the ZO between the years 1911 and 1920. Warburg's involvement, instigation, and management of almost all the Zionist expeditions and investigations that took place between 1903 and 1914, alongside his unique emphasis on climate investigation, demonstrate how the transfer of knowledge and experts between different colonial enterprises enhanced Zionists' assimilation into general colonial scientific discourses.[16]

Alongside Zionist absorption of colonial knowledge in Europe and its colonies, Jewish experts were also exposed to colonial climate science via the imperial sponsorship provided to the Zionist enterprise in Palestine. This sponsorship was specifically relevant to the British support of the Zionist settler cause. As Shira Robinson notes, the "empire's sponsorship . . . with the blessing of the League of Nations, had done much to facilitate the development of a Jewish national home."[17] While this sponsorship is often presented in the historical literature as starting with the Balfour Declaration of 1917, it reflected a much older tradition of English Protestant interest in a Jewish national revival, which was expressed between the years 1902 and 1904 through attempts of British Empire officials to find alternatives to Palestine for Jewish colonization.[18]

British support for the "return" of Jews specifically to Palestine was similarly expressed before 1917 with, for instance, the activity of the Palestine Exploration Fund (PEF), established in the 1860s. This organization, like other analogous organizations operating in Palestine during the last decades of the nineteenth century, not only mapped and researched the country's biblical mythology while advocating its revival by the Jewish people—as it officially stated—but also contributed to the production of colonial knowledge concerning the local environment and climate, which it frequently shared and exchanged with Zionist experts in later decades.[19] Moreover, during the Mandate period (1920–48), British sponsorship of the Zionist project was often manifested in official scientific collaborations between Zionist and British individuals and institutions, even if the relationship between the *Yishuv* (the Jewish settler community in Palestine) and the British local government was not always stable and was sometimes even conflicted.

Therefore, as Eitan Bar-Yosef claims, Palestine presents a challenge to the binary logic of Said's orientalism. Rather than being a case of East/West, either/or, Bar-Yosef identifies it as a case of this/that/the other. In other words, unlike other destinations in the East which were perceived as antithetical to the West, the biblical land was ambiguously addressed as both exotic and strangely familiar, terrifying as well as manageable, as this and the other.[20] While this logic has so far pertained mainly to Britain's religious approach to Palestine, it also serves as a central component of the Zionist settler colonial identity in this country.

Thus, for example, while the investigation of warm climates among European colonialists in other destinations often expressed a sense of apprehension that was meant to "protect" colonizers from the so-called enervating influences of these climates, Jewish settlers in Palestine had a much more intricate approach to the local climate. This was a result of Zionist ambiguity towards Palestine and its native population. On the one hand, Jewish Europeans wished to view themselves as Occidentals settling in the Orient. On the other, they hoped to establish themselves in the country and become native to it. Thus, while they too feared the impact of warm climates on their lives, Jewish settlers were sometimes open to studying indigenous Palestinian methods of coping with the local climate, while simultaneously usually adhering to Western technological solutions. In so doing, similar to other contemporary settler societies, Jewish experts focused on how Jews could best *acclimatize*. In other words, while the colonial approaches to the study of climate in Palestine, mentioned above, mirror the knowledge absorbed by Jewish experts in metropolitan and imperial contexts, many Jewish experts who will be discussed in this book specifically expressed settler-colonial perspectives towards the study of the local climate in Palestine.[21]

Such perspectives are manifested, for instance, in the biography of the renowned Jewish agronomist Aaron Aaronsohn. Born in Romania in 1876, Aaronsohn immigrated to Palestine with his family at the age of twelve, following the economic and social hardship experienced by many Jewish communities in Eastern Europe in the last decades of the nineteenth century. As a youth he worked on the agricultural farms of Baron Edmund de Rothschild, which were often run by French agronomists, and at the age of eighteen he went to study agronomy in the École nationale supérieure d'Agronomie de Grignon, where he became acquainted with the French settler-colonial agricultural experience in Algeria. Following his return to Palestine, during one of his excursions to Mount Hermon in 1906, Aaronsohn discovered emmer (*Triticum dicoccoides*), which was believed to be 'the mother of wheat'. This discovery brought him great professional fame, and Aaronsohn was consequently invited to lecture about his findings around the world, thus, creating many professional connections with international experts, mostly in settler societies.[22]

Growing up in Palestine before the advance of the interwar Zionist mass immigration, Aaronsohn was fluent in Arabic and was interested in understanding Palestinian traditional agricultural practices. In 1910, he established an agricultural experimentation farm in Atlit. This station focused mainly on growing various local types of grain including wheat and barley, as well as local fruit trees such as citrus, olive, carob, pomegranate, and fig. Moreover, Aaronsohn investigated Arab cultivation methods, such as dry farming, that suited the country's water regime. However, at the same time, he frequently pointed out the supposed backwardness and inefficiency of such methods and suggested improving them with Western cultivation methods.[23]

Aaronsohn's biography largely embodies the Zionist settler-colonial experience. Unlike imperial-colonial administrations, which are often presented in the historical literature as primarily motivated by maximizing profit via extracting natural resources and exploiting indigenous labor, settler societies are usually characterized by populations that move by circumstance or necessity and tend to place great importance on the territory as a desired site of social transformation that they intend to make their own. Accordingly, settlers wish less to govern indigenous people or recruit them in their economic undertakings, than to seize their land, eliminate them or push them beyond an ever-expanding frontier.[24]

It is important to stress that the elimination of an indigenous presence by settler societies is not always expressed in physical annihilation. Before 1948, there were two other prominent forms of elimination in Palestine that we will frequently encounter in this book. The first form of elimination was manifested in formal and informal political, cultural, economic, and physical segregation and separation between Jews and Palestinians. This approach was often linked with feelings of anxiety and disgust "inspired" by the indigenous residents and their lifestyle, and it tended also to associate the local climate and environment with so-called Palestinian neglect. The second form of indigenous elimination, which might initially appear contradictory to the first form, was expressed in the active absorption, fascination with, and even appropriation of local customs, knowledge, and culture. As Lorenzo Veracini writes "a settler colonial project is ultimately successful only when it *extinguishes* itself—that is, when the settlers cease to be defined as such and become 'natives,' and their position becomes normalized."[25] Thus, following the work of Rayna Green, I argue that the cultural appropriation of indigeneity is based on a logic in which non-native peoples imagine themselves as the rightful inheritors of what previously belonged to the local population, thus entitling them to ownership of the land.[26] Within the realm of climate investigation, these two attitudes of rejection and appropriation frequently manifested in Zionist experts' oscillation between expressing sincere interest in studying indigenous Arab methods of coping with the local climate, while simultaneously rejecting them as unsuitable for the living standards of Jewish Europeans and stressing the usefulness of modern science and technology in achieving the desired results.

In addition to expressing settler-colonial approaches to the land (including its climate and environment) and its indigenous people, as explained above, Jewish experts often also consciously compared the climate in Palestine and its different variables with the climate in other settler countries such as California in the United States, Queensland in Australia, and French Tunisia, and aspired to learn directly from these places how to overcome the hazards of warm climates. Thus, this study argues that Jewish experts not only shared similar climates with American, Australian, and French settlers but also saw themselves as sharing similar experiences, such as demographic and territorial interests and concerns, as well as viewing the new territory as a promised land.

The numerous Zionist approaches to the local climate and environment in Palestine also included memories of and comparisons with the climates in European countries that the Jewish settlers had left behind. Such comparisons, among other things, mirrored the difficulties of Jewish European settlers in adjusting and acclimatizing to their new environment in Palestine, and they similarly align with common settler-colonial perspectives about their new territories. As some have recently argued, the historiography of Zionism for many decades focused on praising the movement's so-called successes in Palestine. Studies that have criticized this literature in the last few decades[27] argue that it has given disproportionate attention to a relatively small group of idealist Jewish settlers, identified in the Zionist ethos as "the pioneers" (*ha-chalutzim*) who were believed to be "the paradigmatic Zionists." However, despite the popularity of this group in Zionist historiography, it did not represent the attitude of most of the Jewish settlers who arrived in Palestine during the first half of the twentieth century. The majority of Jewish settlers were, in fact, urban and middle class, and were not necessarily driven by a strong Zionist ideology but rather by anti-Semitic persecution and economic hardship.[28] Accordingly, they often expressed feelings of alienation from and frustration toward the local climate, which, as we shall see, grew in proportion with the refugee crisis in Europe during the interwar period and World War II.[29] During the Mandate period the Jewish community in Palestine increased its demography by 8.4 times (73 percent) as a result of mass immigration. This was also the time when the Jewish economic sector grew larger than the Arab economic sector in the country for the first time.[30] The flourishing sense of Jewish confidence that was accompanied by these changes, alongside the existing difficulties of the settlers in managing the local natural conditions and the escalating political conflict with the local Arab population, similarly influenced the scientific discourse concerning climate, especially during the 1930s and 1940s.

In recent decades many historical and anthropological scholarly works have critically analyzed the cultural aspects of the Zionist transformation of the Palestinian landscape that were aimed at "blooming the desert."[31] The purpose of this study is to focus on the scientific investigation of the natural environment by experts, as well as to place Zionist professional and popular discourses concerning the warm climate in a larger global and colonial context by focusing on the professional links and connections between Zionist and other western experts on a global scale, as well as on the transfer of knowledge concerning climate within Western colonizing or "modernizing" networks. In addition, influenced by the unique trajectory of Jewish history in general and of modern Jewish history in particular, scholarship on Zionism tended for many decades to stress the Zionist movement's exceptionalism. The intention of this book is different. As stated above, this study aims to juxtapose the Zionist project to other colonial projects, and to present it in relation to, and correlation with, other similar approaches toward warm climates.

Nevertheless, emphasizing the similarities of Zionism to other colonial enterprises does not mean that this colonial project did not have any unique features. One of the most obvious traits of the Zionist colonial project in Palestine was the aforementioned liminality of its members between east and west. While in Europe Jews perceived themselves—and, indeed, were perceived by others—as an oriental minority in the Occident, their attempts to obtain a non-European territory led them to see themselves as an occidental minority in the Orient. While Jews indeed adopted an occidental perspective about Palestine and its indigenous people, they have often been presented as Orientals themselves whenever occidentals talked about or imagined the East.[32]

This condition of in-betweenness can also be seen in the shifting Zionist approaches to the environment in Palestine, which were not always consistent. As we shall see, the aim of establishing a settler nation of European Jews was, at first, based on a romantic belief in their autochthonous belonging to the Land of Israel. Thus, at the turn of the twentieth century, many Zionist leaders and thinkers attempted to highlight what they perceived as an organic link between the Jewish people and the environmental conditions of Palestine. However, following the actual encounter of Jewish settlers with the natural realities of Palestine, discussions on climate gradually lost their romantic attributes and instead became associated with growing fear of the perceived dangerous implications of non-temperate climates on Jewish European bodies and minds. While Jewish attitudes towards the climate shifted following the colonization of Palestine, this study does not aim to present this shift as an uncomplicated clear-cut event. Nor does it pretend to situate this change in a specific point in time or to associate it with a particular political faction within the Zionist movement. Instead, its emphasis is on the intricacy and ambiguity of the settlers' physical dislocation, which occurred at different times for different actors.

Another unique characteristic of the Zionist colonial project, which was also present in the various scientific discourses on climate in Palestine, was the role of the Hebrew Bible in the construction of its national mythology. Descriptions of the Land of Israel in religious texts prevailed in both Jewish and in Christian traditions for many centuries. As we shall see in the following chapters, the discrepancy between the imagined biblical climate of "the land of milk and honey" and Palestine's disappointing present-day reality became not only a cultural reference but also an important issue of scientific examination. As Billie Melman and Tamar Novick individually point out, religious sentiments in Zionist science and technology were not replaced by, but instead fused with, modernist sentiments.[33]

Biblical sentiments existed as foundational myths in the development of many national movements, especially those of the settler-colonial type.

Historians claim that notions of "ethnic election" and "redemption," for instance, were shared by many Puritan settlements in Northern Ireland and North America in the seventeenth century and can also be traced in late nineteenth and early twentieth century Afrikaner thought.[34] However, while other nations used the Bible and its names and places in a metaphoric fashion, Zionists aimed to use them in a literal manner, or as Amnon Raz-Krakotzkin describes it, "what makes Zionism unique is that its national consciousness presents itself as a direct interpretation of the Judeo-Christian theological myth."[35] This literal interpretation was also, in part, a result of the growing reputation of biblical studies among mid-nineteenth-century European intellectuals, in combination with the popularity of archaeological surveys of ancient civilizations that focused on tracing ancient sites and evidence of biblical events in the Near East, especially in Palestine.[36]

Throughout most of Jewish history, rabbinical literature was often more central to Jewish tradition and culture than the Hebrew Bible itself. However, during the second half of the nineteenth century, the Bible became *the* formative book for the construction of a Jewish cultural and national identity. It was at this point that Jews began to regard the Bible "as their most important asset and heritage, as a shared foundation of values and world view, and as their great contribution to human-kind."[37] The Zionist utilization of the Hebrew Bible as an indispensable part of Jewish history, and the harnessing of its stories and myths for the narration of a Jewish national past, was presented as a secular search for the nation's cultural sources, which were also meant to unify Jews from different countries and diverse cultural backgrounds and enhance their historical (and at times biological and environmental) connection to the territory that was identified as the ancient Land of Israel. Moreover, as we shall see in the following chapters, the references to the mythological history of the Jewish people in Palestine were also often used by experts to justify the appropriation of indigenous solutions for coping with the local climate.

One of the reasons to investigate human interactions with climate relates to its evasiveness, which can mark it as unnoticeable and therefore perhaps a more candid form of cultural and political expression of power-knowledge. Despite reflecting Western power structures, during the early twentieth century "nature" in general, and climate in particular, were understood as existing separately from wars and politics. As Davis writes accurately, beyond classic territorial wars in the name of nature, the employment of more-than-human actors in colonial science ensures that "settlers bear no blame for the impacts because they are unfolding in the domain of 'Nature' . . . as if they occur independently of human interventions."[38]

Moreover, climate is a tangible and physical condition that affects daily life in very concrete ways, a reality to which Jewish European experts tried to respond in developing various technological solutions. However, as an environmental element not easily tamed by humans, at other times climate could be addressed only via theoretical advice and recommendations. As a result of the abstract and tangible qualities associated with this natural element, when addressing Zionist encounters with the climate in Palestine, this study aims to present and analyze both the discourses associated with Palestine's climate as well as the material and physical engagements with it.

The main fields of knowledge occupied with the issue of climate in Palestine were meteorology and climatology; medicine and race science; architecture and planning; and agriculture and botany. The significance of climate in these specific fields was a result of two main factors. First, the successful management of climate in these fields of knowledge meant a successful management of health, food, housing, and productivity, which were perceived as indispensable aspects for the establishment of a sustainable Jewish settlement. Second, as already mentioned, these specific fields of expertise and the knowledge that they produced about warm climates were already well developed by European experts who were dealing with unfamiliar environmental and climatic conditions in many of their colonial holdings (especially in the tropics). Thus, it was only natural that Jewish European experts would absorb this existing knowledge and use it for their own needs in Palestine.

This book discusses the study of climate in Palestine from the establishment of the ZO in 1897 to the end of the Mandate Period in 1948. Its four chapters each discuss the evolution of Zionist climate investigation within a different field of expertise. The chapters follow a thematic order which is, among other things, a result of the continuous failure to harness Zionist approaches to the climate in Palestine into a coherent linear logic.

Nevertheless, whenever possible I have tried to organize the chapters and sections in a chronological order. As this summary will make clear, besides examining the production of Jewish-Zionist knowledge on climate in general political, cultural, and social contexts, this study also analyses some material aspects that were intertwined within the study of climate in Palestine such as food production, food consumption, water management, building materials, living standards, and hygiene practices. In so doing, it aspires to follow other scholarly works that integrate a wide range of subjects pertaining to a certain environmental attribute and weave them together into a large network of historical relations.

Chapter 1, "Knowing Climate," examines the professional knowledge produced within Zionist climate science. This chapter discusses the colonial history of climate science while focusing on the Zionist absorption of its axioms, methods, and practices. The first section of this chapter presents the Zionist study of the climate in El Arish in the Sinai Peninsula and in East Africa—two destinations that were considered for Jewish colonization at the turn of the century. This section shows how, before World War I, despite the popularity of climate observation, the interpretation of its data into useful information presented a significant challenge for Western experts across the globe, meaning that it had relatively little impact on practical decisions made by the ZO. Instead, its significance during this period was in presenting the ZO as an entity that possessed the required knowledge to pursue colonization in the first place.

It was only after the war that climate science turned into a truly pertinent activity for the *Yishuv*. This development was also related to the introduction of new military technologies as well as to the close collaboration between Zionist and British institutions and experts. Thus, the second section of this chapter shows how general developments within climate investigation transformed the study of climate in Palestine and made it not only more practical but also a tool in the Zionist segregation policy towards the country's Arab population. The transformation of Zionist climate research in the first half of the twentieth century also reflected the gradual consolidation of this settlement project. While at the turn of the century the ZO mainly relied on cautious research expeditions to become familiar with the natural conditions of uncharted territories, following World War I it grew sufficient confidence in its settlement project in

Palestine for its executive institutions to begin establishing permanent meteorological stations in the country.

The third section of this chapter focuses on some aspects within the content of climate investigation and demonstrates how this content reflected the settler-colonial rejection of local knowledge. As we shall see, methodologies used by Jewish experts for climate investigation during the first half of the twentieth century included very few local sources of knowledge about climate and, moreover, rarely compare the climate in Palestine with the climate of neighboring countries and regions. On the contrary, Palestine's climate was more often compared with the European climates with which Jewish experts were more familiar. This section also includes a partial portrait of Palestinian knowledge developed about the local climate (specifically among farmers and their communities), which generally reflected different views and concerns to those obtained by Zionists.

Chapter 2, "Climate and the Jewish European Body," follows notions of climate within the disciplines of race sciences and medicine. The first section of this chapter explores the Jewish discourse on climate and race at the turn of the twentieth century, which was intertwined with general popular and professional discussions on the ability of colonizers to acclimatize in non-temperate regions, as well as specifically with contemporary popularized and politicized notions on Jewish degeneration in Europe and their potential regeneration in Palestine.

Whereas the first section of this chapter focuses on ideas developed mainly in Europe at the turn of the century, the second addresses the actual encounter of Jewish settlers with the natural realities of Palestine, which was often characterized by feelings of anxiety and alienation. During the Ottoman period, Jewish settlers in Palestine expressed traditional medical views on the malign influence of "bad" air and warm climates on human health, rooted in both Western and in local Middle Eastern medical traditions and related to miasma theory.

On the contrary, tropical medicine, officially introduced in Palestine following the commencement of the British Mandate, was supposed to fight old notions that linked climate and health, and instead to explain them via the transmission of parasites and bacteria. Nevertheless, as this section demonstrates, tropical medicine in Palestine often continued to emphasize the role of climate in the generation of disease, albeit in a new scientific manner. Tropical medicine—developed alongside imperial expansions into tropical (and other non-Western) regions—was initially meant to protect Western officials who were sojourning in warm regions from their so-called enervating influences. Nevertheless, as mentioned earlier, Jewish settlers in Palestine wished not only to protect themselves from the local climate but also to take root in the country and indeed to acclimatize to it. Therefore, although they often expressed fear of the moral and physical implications of the Palestinian climate, Jewish medical doctors also promoted the theory and practice associated with another medical subfield, often referred to in the historical literature as medical climatology.

Medical climatology concerned the various physical and psychological effects of general weather factors on human beings. In their research, Jewish physicians in Palestine conducted experiments that were designed to examine if and how the local climate would affect the physical and mental nature of the Jewish body. At the same time, they aimed to find solutions to counter the perceived ill effects of Palestine's climate. These solutions mainly concerned hygienic practices and were designed to instruct the population on how to live, dress, and eat. The third section of this chapter thus traces the colonial and metropolitan sources of medical climatology and describes how this medical branch was manifested in Palestine. Unlike tropical medicine, which required the transformation of the natural environment itself for the so-called improvement of human health, medical climatology was directed at the settlers' lifestyle and at times required them to transform it to better suit the local natural environment.

While Palestinian Arabs were not usually a target group for studies in medical climatology (as it was assumed that this population did not need to acclimatize to its environment), they nevertheless provided an important reference point according to which Jewish settlers tended to position themselves. When wishing to transform the Jewish settler society into a native society in Palestine, Jewish physicians often pointed out Arab behaviors and lifestyle and argued that some of these behaviors should be adopted and copied. At other times, when the settlers wished to highlight their separation from the local Arab community, they presented the local Palestinian lifestyle as backward.

One major concern for physicians and patients alike in Palestine was the *hamsin*. This climatic phenomenon brought dry, hot, and sandy desert winds, believed to cause various physical and mental pathologies. When discussing this phenomenon, several experts suggested the use of air conditioning technologies to help alleviate the heat in the houses of Jewish European settlers. At the beginning of the twentieth century, air conditioning was still regarded as a utopian idea. However, during the 1930s and 1940s similar solutions, such as ice boxes and refrigerators, began to appear in Jewish households in Palestine, mainly for the purpose of food preservation. This is also the theme of the first section in chapter 3, "Warm Palestinian Climate—Cool Jewish Spaces."

The remainder of chapter 3 discusses further technological attempts to overcome the heat in public and private spaces. The second section focuses on urban planning in the early twentieth century, especially the "garden city," which was meant to secure ventilation between houses. The third and fourth sections of this chapter concentrate on climate considerations in residential spaces. Architects of the 1910s and 1920s studied and, at times, copied various local indigenous architectural features as part of their intention to alleviate the temperature in the residential houses of Jewish settlers. Their successors in the 1930s chose instead to study the local wind and sun directions and tried to adapt Western architectural methods to their observations on climate. Although these architects often praised Arab building methods for their resilience to the local climate, they nevertheless consciously tended towards the rejection of these methods.

In the fifth section of the chapter, I analyze roof shapes and the quality of building materials as case studies for changes in the architectural discourse relating to climate. The history of the roof in Palestine demonstrates the uncertainty that prevailed with regard to its climatic role versus its cultural representations. Moreover, in the case of building materials, despite their importance in determining the microclimate of interior spaces, the choice of material was not always influenced by climatic considerations. As I explain, more often than not the choice of building materials was affected by economic, cultural, and national impetuses. Nevertheless, this choice had subsequent implications on the microclimates and the physical comfort of Jewish dwellers in their houses.

Chapter 4, "Climate and the Study of Plants," presents the ways in which Zionist agricultural and forestry experts understood the local climate in

Palestine while focusing on their strategies to both "fix" the climate and to plant "foreign" cash crops that would be affected as little as possible by some of its undesired characteristics (such as its low precipitation levels). The chapter's first section, "Forestation Against Desiccation," presents contemporary theories of climate change and demonstrates how planting trees in Palestine was influenced by colonial aspirations to transform the natural conditions in warm regions, as well as to prevent their so-called cultural regression. In Palestine, however, the link between climate, forests, and civilization received an additional local interpretation. Improving climate through forestation was also related to the Edenic image of the country in the Judeo-Christian tradition which, according to the British and the Zionists, needed to be restored via the Jewish settlement project.

The second section of this chapter, "Acclimatization of Foreign Plants," focuses on the transfer of forestry and agricultural ideas and practices, as well as the transportation of seeds and plants between different colonial or "modernizing" contexts and their reception in Palestine. This chapter also addresses the European and colonial professional training of many Jewish experts, who were later also influenced by the policies of the British administration in the country. One of the major sources of foreign inspiration for Jewish experts in Palestine was the body of knowledge on plant acclimatization, which had been accumulated during recent decades in other settler societies.

Zionists believed that the successful acclimatization of foreign crops required knowledge and experience. More importantly, however, they thought that it required the use of advanced technologies that made plants and crops less vulnerable to local natural constraints. Thus, the third section in chapter 4 reads the irrigation solutions that were introduced into Palestine during the first half of the twentieth century as a response to its climatic challenges and stresses the important role of irrigation in enhancing the materially and conceptually imbalanced power relations between Jews and Arabs in Palestine as reflected in their access to, and utilization of, water.

The last section of this chapter focuses on agricultural experimentation with both local and "foreign" crops by Jewish settlers in Palestine. This section mainly examines how climate investigation, as well as irrigation technologies, influenced the cultivation of fruit cash crops. As this section aims to demonstrate, overcoming natural constraints was not only meant to supply food for the local Jewish community and boost Zionist independence and separation from the local Palestinian agriculture, but it was often also developed for economic purposes beyond the local market, such as the Zionist wish to create an export economy to Europe.

1. Knowing Climate

In 1897, the Zionist Organization (ZO) was established by the Austrian-Jewish journalist, playwright, and political activist Theodor Herzl. One of its central goals during its early years was to find a geographical solution for the pressing condition of hundreds of thousands of persecuted Eastern European Jews. During this period, all Zionists agreed that the solution to the "Jewish problem" was the establishment of a Jewish homeland, but they were divided on the question of where this homeland should be, how the territory for a desired homeland should be obtained, and what kind of Jewish characteristics it would have.

Between 1897 and 1904, Herzl made numerous diplomatic efforts to obtain a charter from the Ottoman Sultan to establish a Jewish settlement in Palestine. Palestine was chosen because this territory was believed to be the historical homeland of the Jewish people. Throughout the centuries Jews had retained strong spiritual and emotional sentiments towards the Land of Israel and, therefore, it was considered the most suitable destination for a Jewish settlement. However, following continuous difficulties and obstacles in receiving a charter from the Ottoman authorities, Herzl eventually began examining other potential destinations, which at times looked as though they might be easier to obtain.

Two significant destinations were proposed to Herzl at the time by officials of the British Empire. The first territory, suggested in 1902, was El Arish, a region in the northern Sinai Peninsula. The second territory, suggested a year later, was Guas Ngishu—a six-thousand-square-mile plateau between Nairobi and Mau in what is today the Republic of Kenya. The latter was also known as the "Uganda Scheme," or the "East Africa Plan," and it is this geographical area that constitutes the main focus of this section.

Following the British propositions, Zionist commissions were sent to explore both regions. The biographical backgrounds of their members, as well as the research they conducted during the expeditions, demonstrate the beginning of Zionist climate research and the ZO's absorption of colonial scientific trends that were prevalent among Anglo-European research expeditions at the turn of the century.

The ability of Britain to offer the ZO colonization in these territories was related to its recent acquisitions in different parts of the African continent. For instance, the offer of El Arish resulted from the opening of the Suez Canal in 1869, which was followed by the seizing of Egypt from the Ottomans in 1882. These events naturally elevated the strategic importance of El Arish for Zionists, who believed that a British-sponsored territory that was located close to Palestine would serve as a vantage point for the future colonization of Palestine itself.

In February and March 1903, a Zionist commission visited the designated area. This commission included Jewish and non-Jewish experts, the majority of whom had gained their professional experience in other colonial enterprises. Indeed, according to the historian Deborah Neill, one important attribute of colonial research expeditions at the turn of the twentieth century was the diverse and international backgrounds of experts who participated in them. Neill claims that the transfer of knowledge and expertise between different colonial enterprises on a global scale was, in fact, the norm in the period between 1871 and 1914.[1] In other words, although German, British, and French experts supported the expansion of their states and were prone to compete with each other on both professional and national levels, they nevertheless shared a similar education and training, a feeling of belonging to the same European cultural heritage, and a mutual belief in their "civilizing mission" in non-Western regions.

Zionist expeditions were no different in this regard. The head of the El Arish commission, Leopold Kessler, was a German-Jewish mining engineer who had previously worked in Rhodesia (modern Zimbabwe) and South Africa. Other members of the commission included Albert Edward Williamson Goldsmid, the former director general of the Jewish Colonization Association (JCA) in Argentina; Emile Laurent, a Belgian professor at the Agricultural Institute in Gembloux who had already participated in two scientific expeditions to the Congo; Thomas Henry Aylmer Humphreys, a British civil engineer and chief inspector for the Egyptian Survey Department; Oskar Marmorek, an Austrian-Jewish architect and member of the Zionist Action Committee; Hilel Yaffe, a Russian-Jewish physician who had worked in Palestine since the 1890s; and Selig Soskin, a Russian-Jewish agronomist who simultaneously maintained high-level positions in both the German and the Zionist colonial enterprises.

Although the commission investigated the natural conditions of El Arish (including, among other things, its climate) not one of the commission members was an expert in climate sciences. The historians Kristine Harper and Phillipp Lehmann explain that although this field of expertise was relatively developed at the turn of the twentieth century and had reached a sizeable audience, it was introduced as an independent academic discipline only following World War I. This meant that, prior to 1918, experts who researched climate were rarely qualified climatologists or meteorologists. As a result of the absence of an official academic field that would support the scientific conventions of climate investigation, data on climate were often investigated by non-experts and sometimes presented in an unsystematic manner.[2]

The El Arish report, published by the ZO commission members in 1903, outlined the local climate, soil, water, and other natural elements within a chronological framework of separate daily reports. Each began with a general description of the weather conditions ("weather warm, brilliant sunshine;" "weather fine, cloudy at times, almost dead calm in the morning, later on pleasant cool north breeze").[3] Numeric data were usually added at the end of the daily report. These included wind directions, daily minimum and maximum temperatures (in Fahrenheit), air pressure, and the temperatures during the morning, noon, and evening (in Celsius). However, at no point were these measurements elucidated in the report.[4]

Generally, the accumulation of data on climate and its interpretation into useful information presented a challenge for Western scientists across the globe at the turn of the twentieth century. The astronomer George Airy, for instance, wrote in the late nineteenth century that "the observing is out of all proportion to the thinking in meteorology."[5] Around the same time, the physicist and statistician Arthur Schuster suggested at a meeting of the British Association for the Advancement of Science that experts should stop observing for five years and instead work on ascertaining what their existing observations actually meant.[6]

The report of the El Arish commission also illustrates that the climatic measurements of the expedition were not always taken regularly and were occasionally incomplete. When such situations occurred, the writers of the report often provided an explanation for the absence of data. Once, for example, they wrote that the weather at night "was too boisterous to risk the maximum and minimum thermometer being left outside so the minimum temperature was not obtained."[7] On another occasion, it was reported that a Bedouin servant who oversaw the observation accidently broke the barometer. The partial, unorganized, and unanalyzed display of data in the El Arish commission report was not uncommon among colonial scientific expeditions, which generally relied on investigation tools rather than on experts' knowledge. Indeed, the ease in transporting and operating meteorological instruments, which rendered data on climate in real time and were perceived as accurate and objective, gave this field of research a reliable scientific status, even when the reliability of those in charge of its research was questionable. This apparent reliability largely influenced the contemporary prestige of climate investigation, which also gained its importance from the increasing trust in numbers that began in the late eighteenth century and was frequently manifested in statistics—a field of knowledge that became, in itself, deeply intertwined with analyzing climate data. In Prussia, for example, the first meteorological institute was established as a branch of the state statistical office. Furthermore, many important statisticians such as Adolphe Quetelet, Francis Galton, and Arthur Schuster, frequently utilized meteorological data to develop techniques in statistics. One of the main objectives of statistics in climate research was to calculate daily, monthly, and yearly averages of weather elements in a growing scope of territories.[8]

Following Mitchell Hart's argument on the role of numbers in Zionist social sciences, I likewise argue here that Zionism's general growing trust in numbers within the natural sciences (and specifically in the field of climate) was initially meant to present the ZO as a political entity that held the necessary knowledge required for colonization. In other words, early Zionist expeditions to various destinations relied heavily on the collection of numerical data concerning climate even when their experts did not always know how to translate these data into practical knowledge.[9]

One of the most important issues examined by the El Arish commission was the supply of water in the region. The members of the commission made great efforts to search for wells and groundwater and placed special emphasis on presenting the underground depths of these water sources within the report. However, despite its strong emphasis on water, the commission did not possess a rain gauge and did not measure rainfall or dew quantities in the region.

Furthermore, although the area offered to the ZO by Britain was restricted to the region of El Arish, the expedition expanded the region of its investigation to Pelusium in the eastern extremes of the Nile Delta, El Qantara on the western side of the Suez Canal, and other sites along the canal. In the report's conclusions, the commission suggested adding these fertile areas to the territory offered by the British government, as well as constructing a pipe system to transport fresh water from the Nile to additional eastern destinations.[10]

A meeting in Cairo took place shortly after the commission had completed its investigation; attendees included the renowned British civil engineer William Willcocks, who had undertaken the planning of the Aswan Dam only a year earlier, as well as Herzl himself. However, the parties were unable to reach an agreement concerning potential cooperation on the utilization of the Nile water. In his memoir, Hilel Yaffe wrote that it was Willcocks's rejection of the proposal to include the future Jewish colony in the water distribution of the river that caused the negotiations to eventually fail.[11] The British reluctance to expand the water pipe for this purpose was most likely a result of their interest in securing cotton crops in Egypt, which were of utmost importance to the empire's textile industry at the time.

As a result of the failed negotiations between the ZO and the British government on the terms of establishing a Jewish colony in El Arish, the British Protectorate of East Africa was offered as an alternative destination in 1903. This time the British had a vested interest in cooperating with the ZO. In December 1901, the Uganda Railway linking Mombasa to Lake Victoria was completed. The ambitious characteristics of this imperialist project was meant to strengthen Britain's hold in Africa, specifically around the sources of the Nile. The railway consisted of 572 miles of railroad, 43 stations, and 1,200 bridges, and represented an investment of over £5 million. However, it soon became clear that the project was not as profitable as expected, owing to its infrequent usage by local populations and the small number of Europeans who visited or resided in this region. To cover the mounting losses of this expensive project, the British decided to offer this region to other settlers under their patronage.[12]

Indeed, as part of Herzl's diplomatic attempts to obtain a charter for Jewish colonization from one of the Great Powers of the time, Zionists occasionally "advertised" themselves as settlers who could contribute to other colonial projects. The British Zionist (and later leader of the Jewish Territorial Organization) leader Israel Zangwill, for instance, speculated that "if Britain could attract all the Jews of the world to her colonies she would just double their white population. . . . [W]e could create a colony that would be a source of strength, not only to Israel but to the British Empire."[13]

Moreover, the East Africa Plan became even more relevant for the ZO following the notorious Kishinev pogrom in April 1903 (in what is now the capital of Moldova), which left dozens of Jews dead, and hundreds more injured, and during which an abundance of Jewish property was looted and destroyed. At the Sixth Zionist Congress in August 1903, Herzl presented the British proposal as a temporary refuge for Russian Jews in immediate danger. Following this proposal, the congress decided by a vote of 295 to 178 to send a research expedition to examine this territory.

Located south of the equator, the area called Guas Ngishu by contemporaries was generally characterized by a tropical climate. However, the altitude of the plateau made its average temperatures relatively low. As we shall see, the discrepancy between how the tropics were perceived in Europe and the way in which Guas Ngishu was described by experts reveals many existing preconceptions not only about the natural conditions of the tropics themselves, but also about the perceived nature of the Jewish people, as well as the climate that supposedly best suited them.[14]

The ZO's commission left for Mombasa in late December 1904 and returned to Europe at the end of March 1905. Similarly to the El Arish commission, the Zionist expedition to Guas Ngishu included "foreign" experts. The head of this commission, Major Alfred St. Hill Gibbons, was a British Boer War veteran and a well-known explorer. Another member of the commission, Alfred Kaiser, was a Swiss orientalist and an adviser for the Northwest Cameroon Company.[15] The third and last member of the Guas Ngishu expedition commission, Nachum Wilbush, was a Zionist engineer and the son of a Russian-Jewish farmer, who had no experience in colonial excursions and was appointed to the group at the last moment to represent the ZO's interests in the final report.

Like the El Arish report, the Guas Ngishu commission's methods of collecting data on climate and its ways of presenting and interpreting it mirrored common colonial scientific practices as well as the unsystematic manner of this youthful academic discipline. Even more so than in the El Arish report, the data presented in the Guas Ngishu report reflected the discrepancy between the use of advanced, relatively accurate measuring instruments and the insufficient knowledge of colonial experts in the field of climate sciences, along with their lack of suitable methodologies by which to address the data.

Given the relatively large area that needed to be researched within a short time frame, Gibbons decided to divide the group so that each member would investigate a different part of the region alone. This division was also the cause of the subsequent publication of three different reports, and a comparison between them tellingly reveals how climate investigation was practiced, how meteorological and climatological findings were displayed, and what significance different types of data display had on the ways in which climate was understood.

As Eitan Bar-Yosef points out in a different context, the first part of Gibbons's report reflected the Victorian experience of Africa often associated with colonial adventure literature, which tended to depict hunters and explorers facing dangerous situations in the wilderness of the so-called "Dark Continent."[16] In the second part, Gibbons addressed the concrete natural conditions in Guas Ngishu, referring mainly to its climate, soil, and water. Gibbons presented his findings on climate in a table which included four columns: date, minimum and maximum temperature (in Fahrenheit), and altitude. Following these results, Gibbons concluded that the natural conditions of the plateau, and specifically its climate, certainly made it suitable for colonization by Jewish Europeans.[17]

Kaiser's report, on the other hand, was characterized by a rigid scientific style and referred to a larger range of natural elements. He, too, presented a table that summarized the local climatic data and that also included four columns: time of observation (date and hour), place of observation, sea level, and temperature (in Celsius). Following the presentation of the data in the report, Kaiser utilized statistical methods to assert—conversely to Gibbons—that the temperature differences between day and night on the plateau were too extreme for humans, animals, and plants and that "immigration on any considerable scale cannot be advised."[18]

Finally, Wilbush's report was written in a simple yet practical style that emphasized the need for potential engineering projects in Guas Ngishu. It also opened with a four-column table of the climatic conditions: date, important events of the day (including indication of place), temperature (in Fahrenheit), and remarks. All the remarks in Wilbush's table were short descriptions of the climatic conditions (for instance, "hot and sultry").[19] Like Kaiser, Wilbush concluded that owing to the country's allegedly poor natural conditions, Guas Ngishu would be "unsuitable for Jewish or any other European settlement."[20]

Tables were popular forms for displaying climatic data. However, besides being a satisfactory way to store information, the tabulation of data on climate did not necessarily assist in producing accurate or absolute knowledge about the natural conditions of places. As we saw above, the reports of the East Africa expedition demonstrated inconsistency in methods of investigation, and even a lack of standardization in tabularization among the commission members, which led to extremely different interpretations of the same climatic conditions.

As demonstrated, each of the three tables depicted slightly different elements, presented in a different order. All three reports included the date of the observation, but only Kaiser mentioned the specific times at which the measurements were collected. While Kaiser indicated both the place of the observation and its altitude, the other two referred to only one of these elements—Wilbush to places, and Gibbons to altitudes. All three reports recorded the temperature—two in Fahrenheit, and one in Celsius—yet this was the only climatic element measured by the commission. Humidity, precipitation, air pressure, and other commonly measured climatic elements were absent from the report. In addition, although Gibbons displayed the minimum and maximum temperatures, he did not dwell on this distinction, whereas Kaiser, who did discuss the extreme temperature differences between day and night in Guas Ngishu, did not provide similar categories within his chart.

The arguably unprofessional conduct of the commission members was also expressed by the members of the commission themselves. Gibbons wrote in his report:

A study of Mr. Wilbush's report leaves on the mind an impression of supreme disappointment. A young man serving his first apprenticeship in Africa returns after a short six weeks' experience and dogmatises with supercilious self-confidence, not merely on matters within his own limited experiences but on districts and circumstances beyond the scope of his personal observation. I can best describe this report as a result of the crude conjectures of a very limited and unmethodised experience, and cannot recommend that it be taken into serious consideration.[21]

It should be noted that, beyond strict professional disagreements, personal tensions and criticism constituted the dynamic among the committee members of the Guas Ngishu expedition, especially between Gibbons and Wilbush. This tension can also be seen in the climate section of Wilbush's report, where he voiced his own disapproval of Gibbons's findings and his strong objection to British colonial approaches to both climate and indigenous forms of knowledge in non-European geographies. Wilbush wrote:

If the territory offered is really of the beautiful countries of the world, . . . then following the law of the survival of the fittest, it ought to belong to any strongest tribe that might have adapted itself to the country. . . . If a country is uninhabited there are probably natural reasons which make it unsuitable for human habitation, for humanity is actuated by reasons entirely different of those of the English colonial enthusiasm. It does not pine after beautiful climate and happy hunting grounds as the latter seek, but its aspirations are founded on the possibility of eking out an existence. It would be judging very superficially to say that ignorant negroes [sic] who settled in great masses of more than a hundred per square mile (40 per square kilometer) in the unhealthy Victoria Nyanza coast, would reserve the beautiful and healthy highland territory for a state of the future for white Europeans. If we consider not only the climatic question, but all questions of existence, we shall easily come to understand why the populations of the unhealthy coasts die of epidemics, yet fail to emigrate to the highland which is only about twenty to forty miles distant.[22]

Wilbush's rejection of the East Africa Plan is often understood as reflecting a common contemporary political stance within the ZO, which did not assent to considering any other destination for Jewish settlement besides Palestine.[23] Warburg, who oversaw the appointment of the commission members, was, in fact, also among the opponents of the British offer. Warburg wrote that "the idea of sending Jewish workers to tropical plantations is out of question . . . even if the soil and the climate are suitable as the literature says."[24] Consequently, his choice of a Russian-Jewish engineer for the commission—who at the time of the expedition had already invested funds in a factory in Palestine, and was inexperienced in colonial expeditions, entirely unfamiliar with East Africa, and the only commission member who had been asked to volunteer for the task—may have influenced the final conclusion of Wilbush's report.[25]

Not less important, Wilbush's mocking attitude towards British colonial considerations of climate perhaps also stemmed from what he perceived to be the disproportionally large role given to climate in colonial science, and the significant role of scientific data in establishing the power relations that existed between those who possessed knowledge and those who were intended to be subjugated to it. In other words, Wilbush's approach to colonial science can be understood as embodying the liminal condition of fin de siècle Western Jewry, caught between the position of an oppressed minority in Europe, which aimed to liberate itself from a long history of persecution, on the one hand, and yet absorbed colonial ideas and practices as part of its attempt to obtain a non-European territory, on the other. As Ethan Katz, Lisa Leff, and Maud Mandel argue, before settling Palestine Jews taking part in European colonial endeavors often toggled between master and victim positions.[26]

The East Africa Plan presented one of the most disputed episodes during the early years of the Zionist movement, as it highlighted the existing tensions between different Zionist factions. As demonstrated above, climate was utilized as an argument both for and against the Jewish colonization of Guas Ngishu, but this was not simply because temperature, air pressure, humidity, and wind directions were understood as essential elements for colonization. Rather it was used as an argumentative tool because of the socio-scientific status of climate science at the turn of the century.

Eventually, the inessential role of climate in choosing a territory for the Jewish people was evidenced by the fact that the climate in Palestine was not necessarily presented as more favorable than that of Guas Ngishu. Thus, despite the consistent collection and analysis of data on climate in the reports of the East Africa and El Arish expeditions, in the final rejection of these plans Zionists drew on a longstanding tradition of Jewish longing for Zion rather than considering the scientific findings they had initiated. Nevertheless, the frantic collection of climatic data and the heated discussions of such data by Zionist activists highlighted the empirical qualities of this natural element to assist the ZO in its desire to portray itself as a potential government.

As stated earlier, I argue that during the early twentieth century climate research was often used by the ZO to introduce itself as an official entity representing the Jewish national aspiration that also held the "right" knowledge needed for colonization. Nevertheless, its actual gathering of data on climate and the analysis of these data by colonial experts was usually poor in quality and could hardly be used to draw any significant conclusions about the specific territories discussed. Moreover, the study of climate by experts and laypeople who were not climate scientists was enabled by the field's strong dependency on instruments, as well as the high socio-scientific importance that was attributed to the precise quantitative data produced by these instruments. In other words, thermometers, barometers, and rain gauges could be operated by almost anyone provided with basic training in how to use them.

CLIMATE INVESTIGATION IN PALESTINE

Owing to the weakening power of Ottoman rule in the mid-nineteenth century, a period of reformation had begun which included the "capitulation system." The capitulations were a series of bilateral contracts between the Ottoman and European powers that granted the latter rights and privileges in the empire, especially in the spheres of trade and goods. This development marked an increase in Western interventions in Palestine and, among other things, a growing presence of colonial scientific institutions whose aim was to increase the colonial hold of their nations in the region.

Although interest in climatology and meteorology began to penetrate the Ottoman Empire in the late nineteenth century, it was usually mainly practiced in Turkey itself. It is assumed that the first person to conduct modern climate research in Palestine was a Scottish physician named Edgar McGowan in 1846. From the 1860s, most climate investigations in Palestine were conducted by the British Palestine Exploration Fund (PEF) and the Scottish Meteorological Society under the supervision of the physician Thomas Chaplin, director of the British hospital in Jerusalem. Following the activity of the English and the Scottish, other nations began establishing weather observation stations in Palestine during the late nineteenth century. In 1896, for instance, Otto Kersten, a German chemist, geographer, and the General Consul of the German Reich in Jerusalem founded the German Palestine Association (Deutsches Palästina Verein [GPA]). As part of its activity, this association established several weather observation stations in the country.[27]

The official purpose of the PEF, the GPA, and other similar national-scientific associations was usually to map and rediscover the biblical history of the Holy Land. As mentioned earlier, the Hebrew Bible was generally central in Protestant culture, and it was exceptionally important in modern British culture, society, and politics. Barbara Tuchman even claimed that it came to be "the national epic of Britain."[28] However, it should be recalled that the PEF and the GPA also had military, economic, and political interests that stretched far beyond simple historical and religious interests. Britain was interested in Palestine because, among other things, it held a strategic position in securing Britain's route to India, and the PEF served an important role in turning these aspirations into a

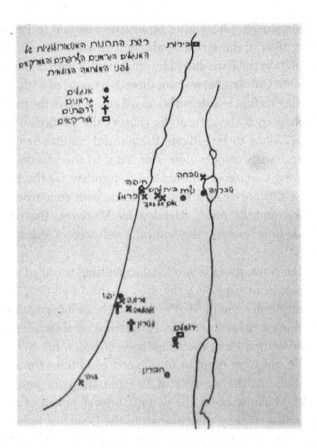

Figure 1: Dov Ashbel, map depicting the international
meteorological stations in Palestine before World War I,
1940. The dots represent British stations, the Xs German
stations, the crosses French stations, and the squares
American stations. Courtesy of the Jewish National Fund
and the Omanut Company.

reality. Thus, for instance, the PEF's cartographical surveys conducted
between the years 1871 and 1878 and 1913 and 1914 were performed
by members of the Corps of Royal Engineers, among them future imperialist agents such as H. H. Kitchener and T. E. Lawrence. Bar-Yosef reminds
us that the 1917 takeover of Palestine by Britain would have been far less
successful had it not been for the detailed maps produced by the PEF in
the preceding decades.[29]

The first known Jewish weather observation station in Palestine was established in 1897 in the agricultural school of Mikve Israel. The school, founded in 1870 by the Paris-based international Jewish organization the Alliance Israélite Universelle, was not directly associated with the activity of the ZO during its first few decades. As we have seen in the previous section of this chapter, at the turn of the century the ZO preferred sending scientific expeditions to investigate the natural conditions of potential territories for Jewish colonization. Adhering to this strategy, in 1903 the Sixth Zionist Congress endowed the Committee for the Exploration of Palestine (CEP) partly as a response to its own concurrent research in East Africa and El Arish. Headed by Warburg, the committee's main goal was to investigate the natural conditions of Palestine and its surroundings.

In 1904, the committee's first official expedition travelled to Palestine with the objective of exploring the area of the Dead Sea, the Judean Desert, and the Jordan Valley. Westerners had been intrigued by this geographical region's great economic potential for several decades as a result of three main factors. First, the abundance of water, the warm climate, and the fertile soil were thought to be good conditions for agricultural enterprises. Second, the streams of the Jordan River and its creeks were understood as useful for the potential development of mechanical energy production. Finally, the minerals of the Dead Sea were beginning to be valued for extraction purposes and could be sold in the global market.[30]

Similar to the other Zionist expeditions just discussed, the CEP expeditions to Palestine emphasized the importance of climate investigation and were also chiefly composed of non-Jewish experts. The 1904 expedition to the Dead Sea, the Judean Desert, and the Jordan Valley, for instance, was led by the German geologist Max Ludwig Paul Blanckenhorn, who was accompanied by the young and—at that time—less experienced Zionist agronomist, Aaron Aaronsohn. Blanckenhorn had already participated in German and Ottoman scientific excursions to the region before being recruited for the job by the ZO. In 1897, he began to work for the Prussian Geological Regional Office (*Preußischen Geologischer Landesanstalt*), and shortly afterwards was elected as head of the GPA.[31]

Following Blanckenhorn's expedition to Palestine, the relationship between the ZO and the GPA grew stronger. An article published in 1909 by Blanckenhorn revealed, for instance, that most of GPA's weather stations in the country were, in fact, funded by the ZO, but adhered to the scientific model of the GPA on Blanckenhorn's advice. This collaboration meant that, once again, Zionists were interested in imbibing climatological and meteorological knowledge from more experienced colonial entities.[32]

As mentioned earlier, the Zionist aim of obtaining a non-European territory for Jewish settlement would not have been possible without European colonization and could not have been accomplished without European imperialism. But while this claim is usually made in reference to the general relationship between Zionist institutions and the British Empire, the cooperation between the CEP and the GPA demonstrates that Zionist reliance on colonial and imperial power and knowledge was not only limited to British sponsorship.

In addition, the significance of the Zionist collaboration with the GPA can also be traced to the fact that after sending several expeditions to examine the natural and social conditions of Palestine, the ZO now finally began investing in permanent meteorological research in Palestine. The very establishment of permanent weather observation stations in foreign territories indicated the transition of colonizers from caution towards a more confident approach to new and unfamiliar territories, as well as their willingness to take risks and invest in these places.

In this context, the professional collaboration between the CEP and the GPA did not only have political significance. It was also related to practical considerations. In an article from 1910, Blanckenhorn reported that the German operators of the old German weather observation stations in the country were not always reliable sources of information because many of them returned to their homeland after completing a short stay in Palestine, and some went on annual leave or became ill and, in some cases, even died. Blanckenhorn argued that this variety of circumstances affected the quality of the scientific data in the stations which were producing inconsistent, and often false, information.

To solve this problem, he suggested establishing new weather observation stations that would be based in existing towns and villages and would

rely on local operators. These operators were to be found in the settlements of the messianic German Temple Society *(die Tempelgesellschaft)*,
such as Wilhelma, Sarona, and Waldheim; in Jewish settlements, such as
Zikhron Ya'akov and Melchamia [Menachemia]; and in Arab villages and
towns, such as Nazareth, Bethlehem, Ma'an, and Jericho.[33] Thus, as the
historian Norton Wise points out, the process of scientific standardization
depended not only on an extensive set of agreements regarding materials,
instruments, methods, and values, but also on the accomplishment of an
extended network of practitioners sharing similar knowledge and
aspirations.[34]

Nevertheless, according to Blanckenhorn, the stations that were located
in Arab towns and villages were harder to operate than others, owing to
what he called "the lack of interest [in climate research] among the
Arabs."[35] Such claims were not exclusive to Arab operators, but rather
reflected a more general gap between the values and beliefs of experts and
those of lay operators. For instance, similar complaints were made by
Aaronsohn in 1905 about the Jewish volunteers who were operating the
weather observation station in Zikhron Ya'akov and later in Atlit.[36]
Avraham Baruch (Rosenstein) who was collecting meteorological data in
the Gimnasya Herzelia school, also often complained about the inefficiency and delays of operators in the Ben Shemen station who were supposed to send him their regular findings.[37]

Following the outbreak of World War I, the cooperation between the
GPA and the ZO was interrupted. As a result, during this period Zionists
increasingly established their own weather observation stations in the
country, mirroring their growing confidence in climate science as well as
in the Zionist settlement project in Palestine. The Jewish meteorologist
Dov Ashbel reported that following the end of World War I every new
Jewish settlement established its own weather observation station. These
stations were usually financed by the Jewish Agency (JA), which acted as
a non-official Zionist government in Palestine.

Moreover, from the 1930s onwards all weather stations were connected and supervised by the Meteorological Department of the newly
founded Hebrew University of Jerusalem, directed by Ashbel himself.
According to Ashbel, the presence of weather stations in Jewish settlements became so prevalent during these years that by the late 1930s a

map of the existing stations was able to delineate the areas of Jewish settlement within the country. This was, according to him, in contrast to the
lack of weather observation stations in Arab towns and villages.[38]

The ability to demarcate areas of Jewish settlement in Palestine via
their scientific and technological activity was not only evident in relation
to weather observation stations. Fredrik Meiton and Shira Pinhas independently demonstrate how the electric grid and a network of roads con-
necting Jewish orchards to the Jaffa and Haifa ports (also known as "the
Orange Belt"), built during the 1920s and 1930s, similarly defined
the borders of the future Jewish state and served as the infrastructure for
the Peel Commission Report.[39]

The area that was supposed to become the Jewish state in the Peel
Commission report was a narrow strip of land along the coastline (between
Jaffa and Haifa). It was nicknamed "the Orange Belt" by contemporaries
due to its abundance of citrus orchards. Between the 1920s and the
1930s, the Orange Belt transformed from an area sparsely populated by
Jews into a Jewish center. During this period the area was also subjected
to Palestine's densest road-paving schemes so that fruits could be transported quickly to markets and ports.[40]

In other words, the increase in weather observation stations during the
1920s and 1930s in this very area was a result, among other things, of the
intensification of Jewish immigration to Palestine, as well as of the consolidation of the economic and political objectives of the Yishuv (which
Meiton terms "technocapitalism") that emphasized the growing necessity
in climate investigation for concrete and practical purposes. Indeed, the
knowledge produced in the meteorological stations of the Jewish settlements in Palestine from the 1920s onwards served general attempts to
comprehend the local climatic conditions that were unfamiliar to the
Jewish settlers. As we shall see in the following chapters, it was mainly
required for the investigation of other fields of expertise, especially agriculture, medicine, and architecture.

As mentioned earlier, the ability to draw concrete conclusions from
meteorological investigations grew significantly after World War I. This
was mainly true in the field of agriculture. During the nineteenth century,
meteorological studies concerning agriculture were primarily occupied
with questions of plant and crop distribution in different colonial settings.

The International Meteorological Organization, founded in 1873, as well as other national meteorological institutions in the West, had already started to discuss the application of meteorology to agriculture and forestry in the last decades of the nineteenth century. However, it took another half-century before these institutions began to produce efficient knowledge on the topic. According to Giuditta Parolini, the famine in many parts of Europe after World War I created a pressing need to build a more resilient agricultural system. The concrete demand for food in conjunction with the growing use of statistical methods to predict yield largely contributed to the importance of meteorological investigation during the Interwar period.[41]

Despite the centrality of weather observation stations in the development of food markets (a theme which will be further explored in chapter 4), their technological advancement during the 1920s and 1930s in Palestine was, simultaneously, a result of the advancement of modern warfare technologies during World War I, and close collaboration between British and Zionist authorities following the establishment of the British Mandate.

Fredrik Nebeker writes that although the weather had always been considered important in the conduct of war, it was not until the twentieth century that meteorological staff became a standard feature in military organizations.[42] The meteorologist Ernest Gold, one of the founders of the British military meteorological service, wrote, for example, that before World War I "the army had little use for meteorology: the attitude of the general staff [was] that the 'army fights its battles with guns and bayonets and not with meteorology.'"[43] Nevertheless, the introduction of chemical weapons and aircraft as major components of the first global war rapidly effected a change in the utilization of meteorological data. Robert A. Millikan, commander of the Meteorological and Areological Service of the Signal Corps during the war, also wrote that "the biggest element in the effectiveness of a modern army is its artillery and . . . the effectiveness of the artillery is dependent entirely upon . . . wind corrections."[44] Indeed, in Palestine, Ashbel reported that the investigation of climate during the war among the British, German and Turkish military was mainly for aviation purposes. According to Ashbel, the most important aspect of their examinations was the "behavior of winds."[45]

By the end of World War I, all major European powers had introduced meteorological services within their militaries and the amount and frequency of reported meteorological information was unprecedented. As we shall see, Zionist climate research in Mandate Palestine likewise owed much of its development to the introduction of these new military technologies—an evolution that also charged the Arab-Jewish conflict with new connotations.

As is widely known, in November 1917 the British government, despite still being at war, announced its support for the establishment of a Jewish homeland in Palestine by issuing the Balfour Declaration. Following Palestine's conquest by the Allied Forces, it was placed under the military rule of the Occupied Enemy Territories Administration, and in the summer of 1920 a civil British administration replaced the military one. This administrative shift also led to fiscal and military adjustments, mainly resulting in a drastic cut in military expenses and manpower as most of the imperial troops were replaced by armored cars and airplanes.[46]

The most important contribution to climate research in Palestine made by the British Empire was its decision to establish the Department for Civil Aviation (DCA) in 1934. This decision was accompanied by another important resolution to build an international airport in Lydda, which would include the first meteorological service in the country. This decision was clearly related to Britain's military and commercial interests in the region and its need to create a convenient stop for airplanes on the long journey to India. It became very handy during World War II when Britain was able to use both the airport and meteorological service for its strategic needs.[47]

During its early days the meteorological service of the DCA consisted mainly of British colonial officers. However, it did not take long for this service to begin cooperating with local Jewish experts and institutions. As we shall see, unlike in other colonial and postcolonial circumstances in which Western powers provided knowledge and technology to non-Western countries, in the case of Zionists (and other settler colonies) there was a two-way transfer. In fact, by the end of the 1930s the DCA meteorological service was headed by the Jewish meteorologist Rudolf Feige. Feige's biography epitomizes not only the cooperation between Jewish experts and the British administration in Palestine in the field of climate science, but also the interrelation between meteorology and aviation, as well as the importance of World War I in advancing both fields. Moreover, it was Feige's familiarity with aviation technologies that meant he was able to migrate to Palestine in the first place and become head of the meteorological service.[48]

Born in Breslau, Germany, in 1889, Feige studied physics and mathematics at his hometown university. During World War I, he volunteered in the Zeppelin Unit of the Kaiser's army, where he served as a meteorologist and a physicist. Following the end of the war, Feige was elected head of the Breslau-Krieten regional meteorological service, which owned two airplanes for scientific use. Nevertheless, following the 1919 treaty of Versailles, flying became strictly limited on German territory, and German official and private institutions were forbidden to continue developing aviation technologies. As a result of these restrictions, German flying enthusiasts turned their attention to gliding, an activity that soon became a patriotic and even rebellious sport in opposition to the Allies. Feige, too, demonstrated great interest in gliding and frequently utilized gliders for meteorological research.[49]

Following National Socialism's rise to power in 1933, Feige was arrested and imprisoned for a year. After his release, he decided to migrate to Palestine. However, during the 1930s and 1940s, Jewish immigration to Palestine was sometimes limited by the British authorities, and Feige was forced to find an alternative way to arrive in the country. In 1935, a few months before the beginning of the second Makabiyah Games (often referred to as the Jewish Olympics), Feige assembled a group of twelve Jewish aviation experts, and together they applied to compete in the games using gliders. This was the first (and last) time gliding served as a competitive category in the Makabiyah. The gliding group included engineers, pilots, technicians and meteorologists, all of whom would become central figures in British as well as Zionist aviation and meteorological activities in subsequent years.[50]

Besides being a practical means by which to bypass the immigration restrictions in Palestine, Feige often advocated the advantages of gliders in the study of meteorology. In a short essay he published in 1935, he explained why this activity was suitable for climate research in general, and particularly in Palestine. He wrote that whereas most countries utilize airplanes to measure temperatures and humidity, gliders, which according to him had not been used enough for systematic climate research, are much more accurate for the estimation of wind direction, as this light aircraft depends entirely on the vertical movement of the air and can help predict the creation of clouds and rain.[51]

After settling in the country, Feige and his German colleagues joined the "Flying Camel" (*Ha-Gamal Ha-Meofef*), a Jewish gliding club founded in 1933, and together with other local gliding clubs stimulated great enthusiasm among the Zionist youth. Similar to the restrictions imposed by the Allies on German aviation, Jews in Palestine were at first not permitted to develop aircraft under Mandate rule. Thus, equipped with technological knowledge from Germany, the Flying Camel and other Zionist gliding clubs invested in this activity while also collecting meteorological data. Academics at the Hebrew University, including Ashbel, also joined the flying efforts and taught courses in engineering, meteorology, and aeronautics to club members. Moreover, in 1935 the Jewish gliding clubs in Palestine were united under the umbrella organization of the Aero Club of Palestine, which in the course of the next decades would become the Air Force of the Hagana and, subsequently, the Israel Defense Forces. This evolution demonstrates, specifically in the Zionist context, the link between the development of military aviation technologies and climate research.[52]

Moreover, it should be stressed that during the 1930s Dov Ashbel, head of the Meteorology Department at the Hebrew University of Jerusalem, and Rudolf Feige, head of the British DCA meteorological service, often collaborated (despite sometimes having personal disagreements), and their institutions frequently shared knowledge. As a letter exchange between officials in these institutions reveals, within this relationship the "Government was to be more or less an agency for gathering data, whereas ... Dr. Ashbel and his department would draw scientific conclusions from this data." Moreover, it was emphasized by local British officials that Ashbel "was to obtain any material he desired and was to be served in every possible way."[53]

The close relationship and exchange of knowledge between Zionists and British officials and institutions in the field of climate science (as well as in its related military activity) might explain why during the 1929 uprisings some local Arabs were reported destroying Jewish weather observation stations in Beer Sheva, Jericho, and Beit Shean. The destruction of Jewish weather observation stations suggests that this scientific field was also perceived by Palestinians as foreign and colonial.[54] Furthermore, the introduction of advanced scientific and military technologies into the meteorological scene rearticulated the increasing segregation that existed between Jews and Arabs in Palestine during the 1920s and 1930s and situated it "beyond the ground." Through the establishment of weather observation stations and the development of military aviation (which depended on meteorological research), the air also became politicized, thus allowing Jewish environmental orientalism and settler colonialism to manifest itself in multiple spatial arenas.

WORKING METHODS IN CLIMATE INVESTIGATION

During the first half of the twentieth century, the content of Jewish meteorology and climatology often reflected the omission and rejection of local culture and knowledge. When conducting climate investigation, Jewish experts usually did not turn to local sources of knowledge, nor did they usually compare the climate of Palestine with that of neighboring countries or regions. As we shall see, they often compared Palestine's climate with the European climates with which they were more familiar.

Such comparisons were made, for instance, in the display of data on local precipitations. Many experts asserted that the amount of annual rainfall in Palestine was, in fact, not very different from that of places in Central and Western Europe, and that it sometimes even exceeded it. The Jewish physician Aaron Sandler wrote that "the average [annual] Palestinian rainfall is 622mm; whereas in Berlin it is 521 and in London 589."[55] Likewise, Ashbel wrote: "Many times, you hear from our brothers who returned from the diaspora [Ha-Galut] a wrong opinion about the rainfall in our country. Many of them think that the amount of rainfall in Warsaw, Moscow, Odessa, Berlin, Paris and London is significantly more abundant than in the Land of Israel," but "this is a big mistake."[56] Ashbel was right. Indeed, at times, the levels of precipitation in Palestine resemble those in the aforementioned places. However, rainfall in most areas in Palestine is generally characterized by short, heavy deluges, and therefore, although precipitation can reach similar levels to temperate regions, it differs in its overall nature and impact on the soil, fauna, flora, and the rest of the environment. In addition, as we shall shortly see, levels of precipitation vary immensely across different parts of the country and can hardly be generalized as reflecting one single climatic unit.

Nevertheless, the comparison of different climatic variables in Palestine with those of European countries displayed the Jewish geographical imagination. Collective sentiments of desired landscape among settler societies have been a widespread subject of investigation among scholars in recent years. The anthropologists Zali Gurevitch and Gideon Aran claim that, while native people tend to maintain "a practical overlap between *the place* in the physical sense ... and *the Place* as a world of meanings, of language, memory and faith," settler societies often experience a breach between the physical *place* and the *Place* of meaning.[57]

Moreover, Caroline Ford reminds us that reflections about the natural environment often demonstrated how settler societies conceived of their past, present, and future. In other words, the way in which the natural world was idealized, avoided, or rejected echoed how it was experienced emotionally and intellectually. Ford writes that perceptions of nature and landscape "were, of course also bound up with changing senses of place and definitions of 'nature' and 'culture.'"[58] She explains the tendency of settlers to long for other landscapes as a result of "environmental nostalgia," and points out that the Greek word *nostalgia* is derived from the word *nostros*, which means "return home," and *algia*, which means "longing."[59] Furthermore, Svetlana Boym claims that nostalgia is a "sentiment of loss and displacement," a "historical emotion" that is linked with modernity and often follows political, social, and economic upheavals.[60]

The Jewish belief that climate in Palestine was comparable with temperate European climates can be understood as mirroring the circumstances, and indeed upheavals, that led European Jews to leave their countries of origin, a transition that was often accompanied by feelings of nostalgia toward this longed-for natural environment. However, by studying the local climate via the framework of other European climates, Jewish experts simultaneously manifested their rejection and avoidance of the local environment and, moreover, their rejection of local knowledge produced about it.

In addition, the comparison of the local climate with that of European countries was meant to demonstrate the environmental potential of Palestine, whose arid climate was attributed to local so-called mismanagement and neglect, and perhaps also to encourage Jewish European refugees to immigrate to Palestine (instead of to other destinations) and to dispel the fears and alienation expressed by those settlers who already resided in the country. These feelings were not directed at the so-called enervating influence of Palestine's environment alone. As we shall see in the following chapters, they sometimes also included the fear of resembling Palestine's native population.

Jewish scientists usually acknowledged the location of Palestine at the intersection between different climatic regions. Indeed, the climatic diversity of this relatively small territory was frequently referred to by contemporaries as one of the land's unique characteristics. Physician Theodor Zlocisti wrote in 1937, for instance, that "for the person who lives in Palestine even the term 'Palestinian Climate' is problematic—of course it belongs to the subtropical zones, but it is composed of a variety of different climates."[61] However, although acknowledging the diversity of Palestine's climate, experts usually continued to compare it with the climate in other distant geographies.

Already in 1873, the British explorer Charles William Wilson wrote: "Owing to the peculiar formation of the country, there is great variety of climate: that of the Lebanon may be compared with that of *the Alps,* that of the Hill Country with *Italy,* and that of the Jordan Valley with the *Tropics.*"[62] Several decades later in 1902 Herzl described this variety of climates in a similar vein in his utopian novel *Altneuland:* "Where in the world will you find a small country like ours with a hot, a temperate, and a cool zone so close together? In the Jordan valley you have *tropical landscapes,* by the coast the soft beauty of the *Riviera,* and not far away the snows of Lebanon, Anti-Lebanon, Hermon, all lying within a few hours of each other by rail."[63] A decade later Sandler wrote: "The climate of the coast is equable, similar to *Southern Italy and Sicily* .... The region of the mountains is more *continental* in the character of its climate. . . . The climate and vegetation of the Jordan Valley are *tropical* in character, and it is not possible for Europeans to live there in the summer."[64]

As we shall see in chapter 4, the comparison of the local sub-climates with other (more "familiar") climates also assisted experts in imagining the profitable cash crops that could be cultivated in certain regions of this country and that could potentially even compete in the global market. The agronomist Elhanan Hershkowitz wrote in this context, "Our country is indeed small, but it is blessed with diverse regions and different combinations of climate and soil, that enable the successful cultivation of most *world fruits* and the supply of fruits throughout the year long."[65]

The comparison of Palestine's climate with other distant climates was not an arbitrary method in Jewish scientific publications. Rather, it was part of a recurrent working pattern. In these publications, the authors first outlined the local climatic conditions, including rainfall, temperature, and air pressure. Then, as we have just seen, they compared these variables with what they understood as similar climatic conditions in other distant regions. Finally, they suggested technological solutions to overcome the local constraints.

Deborah Coen stresses this point specifically in relation to the global economy where the increasing Western interest in climate research starting from the turn of the century was an ironic trend at a time when this economy was becoming more dependent on technologies that could overcome natural constraints. Similarly, while the investigation of climate in Palestine initially aimed to increase Jewish familiarity with the country, the study of the local climate sometimes indeed became a tool to negate local climate as well as the local traditional methods of understanding and coping with it.[66]

What distinguished Zionist knowledge on climate from Palestinian knowledge was largely related to the cultural status which Zionists attributed to different types of experiences of the natural world. The very fact that Zionist knowledge was described as "Western" and "modern" meant a set of specific agreements. According to Latour, within the work of natural scientists "the invisible and the far away is slowly built up from successive layers of amazingly simple perceptive judgments that have to be assembled one after the other with as little gap as possible between every layer."[67] Nonetheless, research has shown that indigenous knowledge is conducted in verry similar ways. John Briggs, for example, reminds us that indigenous "environmental knowledge is constantly acquired, tested and reworked, even if this is only to confirm what is already known. Second-hand information [in indigenous contexts] is not trusted without being first tested and even experienced. . . . [Indigenous knowledge is] based on a combination of observation, experimentation and experience."[68] Thus, the difference between Zionist and Palestinian knowledge on climate was certainly not the accuracy or quality of the accumulated information on this subject. Rather this difference was related to working methods, economic outlooks, diverse cultural traditions and identities, and distinct values and meanings attributed to the local environment that influenced how knowledge was organized, documented, and conveyed.

One of the few historical sources that discusses local Arab perceptions of Palestine's climate is the work of early-twentieth-century Palestinian ethnographers. Following the introduction of modernism into Palestine by the Ottoman authorities—as well as the establishment of Zionist settlements, the commencement of British rule in the country, and the ensuing rise of Palestinian nationalist identity—a group of intellectual Palestinian scholars began to document the folkloric tradition of the local rural culture. This group included, among others, scholars such as Tawfiq Canaan, Khalil Totah, Omar Saleh al-Barghouti, and Stephan Hanna Stephan, who published most of their work in the *Journal of the Palestine Oriental Society*, edited during the 1920s by Canaan.[69]

These Palestinian ethnographers emphasized the importance of climate in the daily life of local Arab peasants. Canaan, who devoted an entire article to the subject of weather observations in Palestinian tradition and agriculture, described at length several local methods to predict the weather. He wrote that on the day of the cross festival (September 14), Palestinian farmers used to exit the village in order to measure the winds as an estimation for the type and power of winds that ought to be expected in the coming winter.[70] Rainbows also served as signs for the predicted amount of future precipitation.[71] Another natural element that Canaan mentioned as indicative of the forecast was the appearance of birds, especially of the common starling. When this bird arrived in large numbers in Palestine during the migration season, local peasants knew that the coming year was going to be rich in precipitation.[72]

Yet, more important was the fact that the study of climate in Palestinian culture was not separated from the life and work of Palestinian *fellahin*, who were most influenced by it. Canaan wrote that:

The daily work leads the *fellah* to regularly observe the sky and the natural phenomena of animals and plants. He discovers that changes in the weather are part of broader natural processes and therefore he transforms his conclusions into poetic forms. Such dictums then pass from mouth to mouth and in each place, they are modified according to the local characteristics . . . and because the farmer knows the weather rules as a result of his daily manifold observation and his annual experience, the content of his dictums reflects the truth.[73]

While Jewish European settlers were mostly concerned with Palestine's summer heat, local Arabs—who were dependent on rainfall for the success of their yield crop—often focused more on winter precipitation in their dictums and customs. A winter rich in precipitation was generally seen as bliss. In this context, Canaan presented seven Arabic expressions which described different types of local rain: "*Naqqatat*—it drips gently, *Rasrasat*—it drips rapidly, *Bahhat*—it rains fast with small drops, *Zahhat*—it rains fast with large drops, *Aburah*—shower falling from a cloud which passes away in a short time, *Za'uq*—the same with heavy downpour, *Sabb* or *kabb min ir-rabb*—very heavy continuous rain (from the Lord)."[74]

Nevertheless, it was not only Jewish settlers who complained about the heat. Palestinians occasionally also complained about the summer heat in July and August. According to Stephan, during the dog days at the end of July and the beginning of August, there were two common sayings: "In July, water boils in a jar" [*fi tammuz btighli 'l'moiji fil-kuz*] and "August flames" [*ab lahhab*]. Stephan added, however, that this heat also brought with it the pleasures of fresh fruits, most of all of grapes, which were celebrated throughout summer.[75]

As in other places, the climate in Palestine was also an important organizing principle in determining the seasons of the year. According to Canaan and Stephan, Palestinians had several ways to divide their year into seasons. The most common division focused on the wet and dry seasons. A division of the year into four seasons existed, albeit following the Palestinian agricultural calendar (which, of course, depended on the local climate). Canaan alluded to a season of sowing and plowing between November and January, the season of harvest and threshing between May and August, the season of grape harvest from mid-July to mid-October, and the season of olive harvest from October to mid-November.[76] Stephan added that besides the four central agricultural seasons, sometimes three other "less important" seasons were mentioned. These included the season of the apricot in the first two weeks of May, the season of the melons from mid-July to the end of September, and the orange season from mid-November until the end of April.[77] Although Jewish European settlers tended to portray Palestine as a barren land, the centrality of the agricultural calendar to Palestinian culture (and its special emphasis on fruit cultivation) demonstrates the importance of farming and crop cultivation for this society. The agricultural calendar also inevitably made the local climate and the ability to understand it a basic requirement in Palestinian life and culture.[78]

As we can learn from the above-mentioned sources, climate was equally as important to local Palestinians as it was to Jewish settlers. However, whereas the latter aimed to decipher climate with the assistance of Western scientific methods, which were largely developed in colonial settings and seemed to them more reliable and accurate, the former relied on accumulated knowledge that served a variety of daily and annual functions, mainly in agrarian contexts but also in determining the local calendar as well as cultural beliefs, and identities.

The local climate dictated farming seasons, which in turn determined collective and individual perceptions of time. For example, climate and agriculture were applied by Palestinians in reference to historical events. Muslims relied on the lunar calendar to determine their religious holidays and Christians on the solar calendar. However, farmers of both religions rarely used their calendars to refer to personal or historical events. Instead, the agricultural seasons, mentioned above, were often also utilized for this purpose. Canaan provided the example of men returning from a war "in the grape harvest season." According to him, when one wanted to be more precise, they would locate the event in the beginning, middle, or end of these seasons (for example, "the son of the neighbors was born in the beginning of the olive harvest season").[79]

In addition to the agricultural seasons, a few other climate-based divisions of the year prevailed among Palestinian farmers. Summer, for example, was divided into three periods: the beginning of summer when temperatures begin to rise from May to mid-June; the time of extreme heat from the second half of June until the first half of August; and the gradual decrease of temperature between mid-August and mid-October. These seasons were not determined by fixed dates. Instead, they largely depended on the changing weather, which also served as an indicator for future weather events. In April, Palestine witnesses its last rains that are supposed to give the earth its final water portion before the beginning of a long summer. Palestinian farmers believed that the prosperity of their local crops depended on these late rains, and when they failed to arrive one knew that a drought was to be expected in the coming year (especially in the more arid parts of the country).

In this chapter, I have highlighted the main underlying ideas and concepts that were associated with the study of climate in Palestine. During the first half of the twentieth century the Zionist study of climate shifted in accordance with the gradual consolidation of the Jewish settlement project. At the turn of the twentieth century, while Zionists were making efforts to find concrete political solutions for the "Jewish problem" in the Russian Empire, they often adopted the strategies that were most available to them in their metropolitan environment and used colonial approaches such as scientific expeditions to study the climate in foreign territories. At the beginning of the century these approaches reflected the insecurity and hesitance of the ZO in relation to various potential destinations considered for Jewish colonization.

General developments in climate sciences following World War I, the commencement of the British Mandate in Palestine, and the solidification of the Yishuv led to the creation of a permanent network of Jewish weather observation stations in the country and enhanced the practicality associated with the study of climate. Nevertheless, climate science in Palestine remained an activity that was sponsored by colonial powers and that served its common objectives. While Jewish climate science imbibed British and German knowledge and expertise during this period, it also served the Jewish community in imagining a different climate with which the settlers were more familiar, delineating Zionist settlements in the country and separating itself geographically, culturally, and politically from the local Arab population.

Although some Palestinian Arabs worked as operators in German and British colonial weather stations, when producing knowledge about the local weather most Palestinians (especially farmers) did not seem to rely much on Western knowledge. The German Jewish physician Fritz Kahn later gave expression to the remoteness of scientific forms of climate data display in contrast to the "real" concerns that farmers faced. He wrote:

> Numbers and curves make too great of an impression. . . . [Statistics] are indeed uncontroversial and yet—they tend to be wrong. . . . Nature does not discuss a 'state of health', the 'average age', the middle temperature, isotherms, and any other types of climatic terms. These terms exist only in the minds of scientists. If you tell a farmer—and a good farmer knows the meaning of weather—. . . 'your farmyard has the median temperature of 16 degrees,' he will laugh at you because *he* knows—what scholars and scientists often forget—that a middle temperature does not exist.[80]

2. Climate and the Jewish European Body

CLIMATE, RACE AND THE "ORIGIN" OF THE JEWISH PEOPLE

Since antiquity, climate has been routinely invoked in the explanation of human moral and physical variations. Following the encounter of Europeans with unfamiliar environments in the so-called New World, climate received renewed attention and gradually became a central explanation for the fundamental Otherness of non-European environments and people.[1] The modern version of the conjecture between race theories and neo-Hippocratic[2] theories during the nineteenth and twentieth centuries was often titled medical climatology. This medical branch, which will be discussed in detail in the third section of this chapter, concerned the various physical and psychological effects of general climatic and weather factors on human beings. Among the most important questions raised by medical experts in this context was whether European people could acclimatize to natural environments disparate from those to which they were accustomed. The notion of acclimatization had clear implications for colonial strategies, in particular determining whether Europeans could or should settle the distant geographies they controlled. Once new territories were settled, settlers began to ask themselves how foreign natural environments, and especially their climates, would physically and mentally affect colonizers in the short and long term.

According to Mark Harrison, before 1800 there was general optimism regarding the prospects of such acclimatization. However, toward the beginning of the nineteenth century this optimism began to fade, a shift that Harrison links to changes in ideas about race, which emphasized the role of heredity over the environment, as well as to the consolidation of colonial rule in many parts of the non-temperate world. Nonetheless, during this time, the concept of acclimatization did not disappear but rather transformed into what Harrison calls "weak transmutation." This view accepted and acknowledged the effects of climate, but drew these effects out over many generations, thereby reconceptualizing climate as "a remote rather than an immediate influence on human development."[3]

Generally, conceptions of race did not preclude the scientific education of emancipated Jews in Europe. On the one hand, Jewish interest in race sciences reflected their successful integration in Western scientific circles. On the other hand, it pointed to the limitations of their integration, while serving as a response to the ongoing denigration of Jews as a so-called "inferior" race.[4] Race was also invoked to reinforce a conception of Jews as a *Volk* and nation instead of a religion. In fact, according to Mitchell Hart, the notion of Jewish racial decay and the rise of Jewish nationalism were deeply intertwined. For Zionists, racial notions were not only a means of strengthening Jewish national identity, but also a tool for clarifying the urgency of the nationalist cause. According to Zionists, Jews were a nation and race essentially different from other races and nations with whom they had integrated and assimilated over the past centuries. Thus, a Jewish homeland would contribute to their racial re-isolation, thereby preventing their perceived degeneration.[5]

However, Jewish scientists, including those who adopted the notion of "pure races," usually preferred scholarly accounts that emphasized the influences of environmental, historical, and cultural elements in the development of races, while rejecting those that focused on heredity alone. By locating Jewish degeneration in environmental conditions rather than in strict hereditary, Jewish doctors and social scientists presented the task of environmental transportation as equal to physical improvement. Jewish scientists such as Elias Auerbach, Arthur Ruppin, and Redcliffe Nathan Salaman explained that the Jewish *Urtypus* (primal type) was created in Palestine in ancient times and thus his "return" to the "biblical homeland" would correspond with the race's ancient nature.[6]

The use of environmental arguments within the contemporary racial and medical discourse corresponded with broader intellectual discussions on the origin of nations and was, in fact, not unique to Zionist thought. The renowned British historian Henry Thomas Buckle, for instance, asserted in the introduction to his *History of Civilisation in England* (1857) that there was a strong relation between the characteristics of civilizations and the environmental conditions in which they were initially created.[7] This notion, which became increasingly prominent in Europe during the late nineteenth century, was also manifested in the work of German geographer and ethnographer Friedrich Ratzel, who in 1897 presented his infamous concept of *Lebensraum* (living-space), which originally concerned the relationship between human groups and the spatial units in which they historically developed.[8]

Arguments about the "natural" link between people and places also entered the political sphere in Zionist discussions of Palestine. Following the tempestuous debate over the East Africa Plan, the Zionist thinker and leader Zeev Vladimir Jabotinsky explained in his 1905 *Zionism and the Land of Israel* why Palestine was the only logical destination for a Jewish settlement.

> The slow and constant pressure of the natural environment—the landscape, the climate, the flora, and the winds of the homeland—is what determines the structure of the nation's soul. . . . Before we arrived in the Land of Israel, we were not yet a nation and we did not exist as a people. It was on the soil of the Land of Israel that the Hebrew nation was created [in biblical times]. . . . When conceptualizing monotheism, we breathed the winds of this land. When we fought for our independence, we were surrounded by its air, which nurtured our bodies. . . . The nation of Israel and the Land of Israel—are one. . . . A different climate, a different flora, and other mountains will necessarily distort the body and the soul, which were created by the climate, the fauna, and the mountains of the Land of Israel.[9]

The notion of climate and nationhood reflected in Jabotinsky's writings echoes the above-mentioned perceptions of the role of climate and the environment as primal and remote influences on the development of the Jewish people and was supposed to highlight the nativeness of Jews in Palestine. Jabotinsky was decisive in his wish to portray the choice to colonize Palestine as driven by rationality. He claimed that those who aimed to present this decision as driven by fantastic religious notions should be aware "that the relations between Zionism and Zion are, for us, not only a strong instinct . . . but a justified conclusion of great importance that is based on pure positivist reasons."[10]

Indeed, at the turn of the century Zionist scholars were aiming to prove in scientific manner the Jewish physical link to Palestine. Only a year before Jabotinsky's text was published, Warburg wrote in the scientific journal of the Zionist movement *Altneuland*: "The fact that Jews are able to tolerate the Palestinian climate cannot be doubted; indeed, it is their original homeland [*Heimat*], and even if they have managed by now to also acclimatize in northern climates, the next generation of Jews living in Palestine will soon feel integrated in its climate just as the local Arabs do."[11] By contrast, Warburg believed that the Germans of the Templer community in Palestine were suited only to Nordic climates. In the same text he commented that "all German attempts to move to the South have so far failed."[12] Dana Suffrin suggests that such rhetoric was also meant to convince the Ottomans of the potential productivity of Jews in Palestine. According to this logic, since Jews were considered both Europeans and Orientals, they could become both brothers and conveyers of "civilization."[13]

By expressing these ideas Warburg also mirrored the previously mentioned skepticism regarding the general ability of Europeans to acclimatize in warm climates (especially in the tropics). Criticism of the possibility of Europeans acclimatizing in the so-called torrid zone began emerging in the second half of the nineteenth century as a result of, among other things, the high rates of European mortality in the colonies in previous decades, which had weakened the colonizers' confidence. The fact that European death rate did not seem to wane over time, even among those who were born in the colonies or had lived there for longer periods, was presented by some in the mid-nineteenth century as evidence that acclimatization was, in fact, not possible.[14]

This type of skepticism also played into the hands of contemporary thinkers who objected to colonialism. The fourth president of the Société d'Antropologie, Jean Boudin, who was known for his objection to Napoleon III's plans to settle Algeria, argued against the possibility of French acclimatization to the north African climate in an article he published under the title "On the Non-Cosmopolitanism of Mankind." Similarly, the renowned German scientist Rudolf Virchow, who had the reputation of a progressive republican, manifested his objection to German colonialism by opposing the possibility of German acclimatization to warm climates. In 1885, Virchow published an article entitled "The Acclimatization of Europeans in the Colonies," in which he claimed that the nations of northern Europe had a higher concentration of Aryan blood compared with southern Europe. This diagnosis enabled him to explain why, for instance, the French suffered greater loss of life in the Caribbean than the Spanish. Virchow wrote of Jews that they were the "least Aryan" of all people and, hence, they were the best candidates for colonization.[15]

This view was also accepted by the prominent British race scientist John Beddoe, who asserted that Jews could thrive better in different climates than other Europeans.[16] Virchow and Beddoe did not necessarily intend to argue for Jewish inferiority, although they did contribute to the assumption of acclimatization as a flaw. According to Eric Jennings, "it would be a small step to transform the image of the protean, acculturable, acclimatable Jew ... into that of the cosmopolitan, errant, and declassé Jew contrasted by nineteenth century racialists with the firmly rooted Aryan."[17]

However, such views on the cosmopolitan nature of Jews did not challenge Zionist thinkers who, by promoting colonization in Palestine, justified the rhetoric of a Jewish rootedness in their so-called historical homeland and primal environment. In other words, Jewish colonization and the "return" of Jews to their "homeland" were seen as one. This irony was partly enabled by the Zionist secular interpretation of the Hebrew Bible. As Raz-Krakotzkin reminds us, Zionism was able to declare itself a secular movement only because it understood nationalism as the best interpretation of theology and as a kind of revelation which manifested its "true" modern content.[18]

Warburg's text, quoted above, demonstrates how as part of this paradigm, Palestinian Arabs were likewise sometimes portrayed as people inhabiting their "natural" environment ("the next generation of Jews living in Palestine will soon feel integrated in its climate just as the local

Arabs do").[19] As we shall see in the following sections, this position occasionally presented Arabs as a source of inspiration for Zionist settlers, who were interested in learning their ways of managing with the local climate. At other times, however, local ways of managing with the local climate displayed a threat.

Moreover, following this logic, Sephardi Jews originating from Middle Eastern countries were considered to have a greater potential to acclimatize to Palestine than European Jews. As Raphael Falk points out, many Zionist thinkers viewed Sephardi Jews as the ultimate original Jews and ideal settlers in the so-called ancient homeland.[20] For instance, Sephardi food consumption and diet, topics which will be further discussed later in this chapter, were viewed as more fitting for the local climate than Jewish European habits. Efrat Gilad points out in this context that while Jewish nutrition experts sometimes saw Palestinian diets as too difficult for Jews to digest, Sephardi diets were believed to be better, because they "bridged" the gap between European Jews and Palestine's climate.[21]

In addition, in relation to the question of labor, Warburg argued in 1909 that the only ones who would be able to cultivate Palestine would be those Jews who fitted its climate. According to him these were Jews from Yemen, Iraq, Syria, and South Arabia: "As laborers they are very desirable since their productivity resembles that of the *fellahin*."[22] A similar assumption was expressed in the 1930s by physician Theodor Zlocisti when he asked, "is the process of adjustment [to the local climate] different among Nordic people, local Semitic people and Semitic people from the diaspora?"[23]

Nevertheless, references to Sephardi acclimatization in the writings of Zionist experts were, in fact, scarce in the period discussed in this book. One potential explanation for the scant attention paid to this group in relation to climate and acclimatization could be that before the 1950s this group was relatively small and less significant from the point of view of the Yishuv institutions that were run mainly by European Jews. Another explanation pertains to their rootedness in Palestine. That is, just as it was assumed that the local Arabs did not need to acclimatize to their homeland, so it seemed that Sephardi Jews, having lived in Palestine and the Middle East for centuries, did not need to go through this kind of so-called physical transformation. As some scholars have demonstrated, a discourse on the racial attributes of Sephardi Jews did prevail in Zionist eugenic discourses; however, these discourses tended to focus less on climate.[24] Accordingly, the role of Sephardi Jews in this study is limited.

Generally, for Europeans during this period productivity and race in warm climates were associated with skin color. This notion had a long history in European racial thought. For instance, On Barak mentions the racial division of labor on steamboats, which was related both to notions on heat and skin color. According to Barak, those who worked in engine rooms of steamboats were usually dark-skinned non-Europeans "who were considered racially suited to the furnace-like temperature of the engine room, which could reach 70 degrees Celsius."[25]

Similar notions were also expressed by Jewish experts in Palestine at the turn of the century. For instance, when Aaronsohn and Soskin embarked on an independent scientific expedition to the Jordan Valley in 1901, the former wrote that in this area "only negros [sic] would be able to survive the burning hot summer days." Furthermore, for Aaronsohn "the question of mortality is frightening. . . . We encounter fertile land beyond imagination, much water for irrigation and it is possible to purchase big territories for 2 francs a dunam. Yet, if we settle Jewish people there, we will sentence them to die."[26]

As we have already seen in the case of the East Africa Plan, when it came to finding the right destination for Jewish colonization, Zionist did not only present Jews as Europeans but also as white people. This was by no means an obvious statement at the turn of the century. Historian Sander Gilman points out that during the late nineteenth century Jews were, in fact, sometimes described as "black," or at least "swarthy."[27] In 1850 Scottish physician and ethnologist Robert Knox referred to "the African character of the Jew." When describing the external characteristics of Jews, Knox wrote that "the whole physiognomy, when swarthy, as it often is, has an African look."[28] Similarly, when visiting North America for the first time, Adam Gurowski, a Polish noble, "took every light-colored mulatto [sic] for a Jew."[29] Likewise, the British–German philosopher Houston Stewart Chamberlain wrote in 1899 that the Jewish race contained African blood.[30]

The view of Jews as black was occasionally also expressed by Zionists themselves, usually in reference to the image of East European Jews. Herzl described Zangwill (whose family originated in the Russian Empire) as being a "long-nosed, Negroid type, with very woolly deep-black hair."[31]

Likewise, literary scholar Axel Stähler demonstrates how, in the context of the East Africa Plan, the Zionist satirical journal *Schlemiel* presented Jews as sharing an affinity with Africans, partly because those Jews who were supposed to settle in East Africa (Eastern European Jews) were regarded as "black" Jews "with beards and sidelocks, garbed in somber black."[32]

If Jews were considered dark at the turn of the century, why were they described as white at the same time? And how, during a period of hardening racial categories, were they able to shift from one to another? Describing Jews as black or, alternately, as white obviously meant positioning them as either inferior or superior. Homi Bhabha claims that the discourse on whiteness concerns authority more than an authentic or essential identity.[33] Indeed, postcolonial thought has led scholars of Zionism in the last decade to argue that by colonizing non-European territories, Zionists gradually began to think of themselves as Western and as white people. Historian Daniel Boyarin asserts that it was modern anti-Semitism that caused assimilated Jewish thinkers to conclude that the only way to regain Jewish "lost integrity" was by embracing imperialism. Moreover, Boyarin explicitly claims that Herzlian Zionism needed colonialism to transform Jews into white men.[34]

Given that since roughly 1904 Zionist ideology had rejected the idea of including Arab labor in its settler colonial economy, the productivity of Jewish settlers in Palestine was seen as crucial, and therefore, the need to secure Jewish acclimatization in the Palestinian climate became even greater. This aim was central to the work of physicians during the 1930s and 1940s and it will be, therefore, discussed in further detail later in this chapter. But before jumping to this later period, other medical traditions, which were likewise entangled with climate in preceding decades, will be explored in the following section.

TROPICAL MEDICINE: A HEALTHY LAND MAKES A HEALTHY PEOPLE

As we have seen, at the turn of the century some contemporary Zionist thinkers in Europe believed in the regenerative benefits of the Palestinian climate for Jewish bodies. However, the actual encounter of Jewish settlers and physicians with Palestine's climate transformed romantic Zionist views on climate into apprehensions. This new concern was manifest in the Zionist plea to cure the Jews' new habitat itself. While the so-called improvement of climate via the concrete transformation of the environment will also be discussed in the first section of chapter 4 (which focuses on Zionist afforestation in Palestine), this section aims to address the medical theories that intended to eradicate diseases associated with the country's warm climate.

The major disease that Palestine needed to be "cured" of at the beginning of the twentieth century was malaria.[35] This disease and its modernist and colonial connotations have been thoroughly studied by several historians in recent years, most notably by Sandra Sufian in her book *Healing the Land and the Nation*.[36] In this section I wish to complement existing scholarly works by highlighting the role of the local climate in Zionist and British medical discourse concerning malaria.

Before the twentieth century, the majority of Western physicians and laypeople attributed malaria to the inhalation of noxious vapors emitted from swamps. The word malaria comes from the Latin *mal-aria*, "bad air." Influenced by ancient Greek texts, these ideas, which formed part of miasma theory, were also common in the Ottoman Empire, and specifically in the Middle East from the Middle Ages. Despite being disputed and challenged by germ theory in Europe from the 1880s onwards, a belief in miasma theory and the general link between warm climates and disease often prevailed among local practitioners and laypeople in Palestine well into the 1920s, even surviving in the very premise of tropical medicine.[37]

The continuation of old medical ideas pertaining to climate within the new theory of tropical medicine was related, among other things, to a gradual process of knowledge production. This was not only true of Zionist medical views in Palestine. Historiographical accounts of tropical medicine in European colonies often emphasize the continuity, rather than the extinction, of medical ideas relating to warm climates during the first half of the twentieth century.[38] Jennings writes, for instance, that "although parasitology and the French Pasteurian revolution identified the microbe as the new enemy, climate remained an important, if not the most important, pathological agent. Early-twentieth-century Pasteurians found themselves still preaching against the widespread belief that climate was to blame for European fragility in the colonies."[39] Nevertheless, such beliefs concerning the link between illnesses and the natural environment were also persistent because despite being an abstract concept, climate was still understood as more tangible than invisible germs and bacteria. Moreover, although not being the direct cause of diseases, climate did sometimes naturally influence the conditions that enabled the development of some bacteria.

In Palestine, nearly all Jewish agricultural colonies before and during the Mandate period were located in low-lying topographical areas near malaria-infested swamplands, particularly the valleys and the coastal region. These lands were considered cheaper to purchase and, owing to their high rates of malaria infection, their Arab landowners were also more likely to sell them. The outbreak of disease in these specific areas led numerous popular Hebrew texts to identify a link between malaria infections and the local natural environment, with an emphasis on its climate. An analysis of these texts also reveals the ways in which the term "climate" was used at the beginning of the century to refer to different scales of the physical environment.

For instance, one article published in the Labor Zionist newspaper *Ha-Ahdut* in July 1914 condemned the purchase of a swampland from the Arab village Tel-Shamam by Jewish institutions. The author of the article, purportedly quoting local Arabs, described the climate of this area as being "so bad that even birds die when flying above it," adding that "it would take years to improve the climate of Tel-Shamam."[40] Owing to this view, he remarked that without "climatic improvements," settlers would continue dying from what he termed "the green fever." The article ended with morbid advice: "Get out and see the gravestones of Hadera and you will understand how many victims are needed to cure climate."[41]

The motif of birds dying above areas afflicted with noxious vapors seems to have been common in early Zionist writings. An analogous story was told about the purchase of Petach Tikwa in 1878, almost forty years earlier, by Joel Moshe Solomon, Jehoshua Stamber, and David Gutmann. They apparently bought the land despite the absence of birds in the region—an absence thought to indicate its "malign climate."[42] In a similar vein, Sandler wrote in 1912 that "a better acquaintance with the climatic

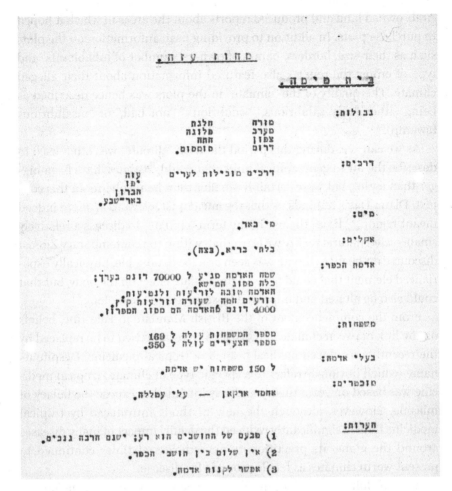

Figure 2: Report of the Israel Land Development Company on the natural and social conditions of Arab-owned lands in Gaza district villages, 1920–1921. Courtesy of the Central Zionist Archives.

state of Palestine would have spared the pioneers many disappointments and saved them from cruel deceptions, and enabled them to make a better choice of the sites where on to found colonies."[43]

Perhaps in response to these past mistakes, throughout the 1920s the Israel Land Development Company (*Ha-Chevra Le-Hakhsharat Ha-Yishuv*) chose to investigate the natural and social conditions of

Arab-owned land and produced reports about the areas in which it hoped to purchase plots. In addition to providing basic information on the plots such as their size, borders, names of owners, number of inhabitants, and type of crops, the reports also featured information about their alleged climate. The quality of the "climate" in the plots was hence described as being either "very salubrious," "salubrious," "not bad," or "insalubrious (swamp)."[44]

As we can see, during this period the term "climate" was often used to describe the air in geographical units that could, at times, have been bigger than a plot, but were certainly smaller than hemispheric. In this context, Diana Davis reminds us that the ancient Greek word *klimata* indeed meant region.[45] From the use of the term "climate" to discuss a relatively small-scaled natural reality we learn that, within the contemporary Zionist discourse on health, climate was seen as a local, tangible, physically experienced element that could have an immediate effect on the body, but that could also be altered and improved by science and technology.

From the commencement of the British Mandate in Palestine, beliefs on the link between climate and diseases were supposed to be replaced by the recently introduced medical branch of tropical medicine. Despite its name (which obviously refers to a specific type of climate) tropical medicine was based on germ theory and was intended to expose the fallacy of miasma. However, although the new methods introduced by tropical medicine indeed significantly reduced the deadly impact of many diseases around the globe, its practitioners nevertheless at times continued to present warm climates as fertile grounds for disease.

As Ryan Johnson reminds us, from its inception, tropical medicine was linked with imperial expansions into tropical regions. Johnson argues that tropical medicine was not just a theory and a practice, but also a scientific language that influenced academics, governments, and industrial interests. One of its main features was to highlight fears of tropical people, places, and diseases at a time when the development of tropical colonies was increasingly linked to the future security and prosperity of the British Empire.[46] As a result of their colonial connotations, germ-based diseases were regarded as tropical diseases even when they broke out in subtropical and even temperate climatic regions. Therefore, Margaret Jones claims that "tropical medicine" at the beginning of the twentieth century was neither consistent nor uniquely tropical, as many measures adopted by this discipline were already used in metropolitan centers. However, despite these similarities, practitioners, politicians, and industrialists' interests maintained that the needs of European bodies were fundamentally different in tropical climates.[47]

Moreover, because Europe saddled warm countries with a great degree of Otherness, the deeper cultural, social, and political meaning of "tropical medicine" was that the tropics denoted a malleable concept, which could be stretched and applied to almost any non-Western locale. Thus, as historian Anne Marie Moulin demonstrates, tropical medicine was implemented, for instance, in North Africa, notwithstanding the fact that this region had markedly different natural and climatic characteristics from the tropics.[48] This discrepancy was also true for Palestine where, as Sandra Sufian has skillfully demonstrated, malaria seems to have been the greatest concern for tropical medicine practitioners. This association of malaria with non-temperate climates can be traced in many official documents. For instance, a pamphlet for schools in Palestine published by the British Department of Health in 1924 declared that "the most striking difference between the health conditions in the Near East and those in more temperate climates is found in the greater prevalence of insect-borne diseases in the former, and of these mosquito-borne disease—malaria—is the most widespread and in this part of the world the most important."[49]

As mentioned earlier, malaria could actually be indirectly influenced by some natural elements which formed the basic conditions for the breeding of mosquitoes. These natural elements usually included a combination of a specific topography, soil quality, rainfall, winds, and temperature, and their study by tropical medicine practitioners from the late 1920s was based on the increasing production of geological and meteorological data during these years, as discussed in the previous chapter. Once particular climatic variables could be measured at different seasons of the year and different times of the day, they could be cross-checked with illness and death rates in specific areas. For instance, the medical literature noted that a certain level of humidity was required for mosquitoes to be effective vectors of malaria. Furthermore, it was argued that mosquitoes were more likely to feed on humans in warm and humid environments, especially after sunset when westerly winds faded, and the air was still.[50]

Another important environmental factor that, according to these studies, caused the proliferation of mosquitoes was the chemical and geological structure of the ground, which determined whether rainwater would penetrate or accumulate on the surface. In the latter case, if water was not directed to rivers and creeks, it naturally transformed into swamps and became the habitat of mosquitoes. Other environmental elements influencing the proliferation of mosquitoes included the amount of water in a basin, its quality, and the distribution and characteristics of the flora and fauna in and around the basin. As one Jewish expert, Israel Jacob Kligler, wrote: "The configuration of the country, the character of the geologic formations, the volume and distribution of rainfall, the temperature and its seasonal variations—these are the physical facts which determine, to a greater or lesser degree, the distribution, prevalence, and intensity of the disease. Of equal importance are the habits, customs, and economic conditions of the inhabitants."[51]

The final point made by Kligler introduced the human factor into the existing nature-disease equation. According to him, poor nutrition increased the risk of malaria infection. He wrote that rural Arabs maintained a poor and uneven diet, and therefore were usually more prone to illness. When food was available in sufficient quantities, however, the local population adopted a diet well-suited to the climatic conditions of the country.[52] Regarding the Jewish population in Palestine, Kligler noted that, although they maintained a more stable diet, their efforts to retain European food habits weakened their immune system. As we shall see in the following section, the link between climate, diet, and disease was central to Jewish medical discourse.[53] Nevertheless, Kligler's text also teaches us that tropical diseases were not only seen as a result of the climate and the environment itself, but also as an outcome of so-called human negligence. The relation between "backward" civilizations and warm and diseased climates was, in fact, dialectic and contradictory. On the one hand, colonialists argued that indolence and negligence were cultural behaviors that flourished in warm climates. On the other hand, these very behaviors were also understood as the reason for "environmental decay."

According to Davis, nineteenth and twentieth century portrayals of the Middle East by Anglo-Europeans represented it as being on the edge of its ecological viability owing to "severe neglect" by the Ottoman authorities and local populations. Davis demonstrates, for instance, how French colonialists in North Africa tended to view themselves as the heirs of the Roman Empire, and to that end aimed to restore the alleged fertility of Algeria, Tunisia, and Morocco under Roman administration.[54] The British government in Palestine drew a similar connection to the Roman heritage of the country. In February 1922, it initiated a cadastral survey in the Beisan and Jordan Valleys. The historical timeline of the survey argued that, during Roman and Byzantine times, extensive canals were dug to keep swamps from forming on the ground. The writers of the survey claimed that it was as a result of the battle between Heraclius and the Muslims in 636 CE that the canals were destroyed, causing the generation of swamps. In other words, the writers associated the beginning of malaria in Palestine with the rise of Islam. In this context, they explain, "there is no evidence at hand to show that any attempt at a definite canalization and drainage scheme was ever tried out [after Roman rule] until the advent of British occupation almost 1,300 years."[55]

This negative narrative about Ottoman rule allowed the British to act in the name of improvement and progress. In addition, although the British were a main force in introducing tropical medical measures into the country, the commissioner, Herbert Samuel, mentioned several other major forces that were involved in fighting malaria in Palestine. These included the International Health Board of the American Rockefeller Foundation, which was involved in the research of tropical medicine on a global scale,[56] the British Malaria Research Unit (MRU), the JCA, and various Jewish agricultural colonies.[57]

As in the case of the meteorological research discussed earlier, British and Zionist (and in this case even American) institutions did not only share operative plans, information, and knowledge. They also shared experts. This trend was epitomized in the biography of Israel Jacob Kligler. Born in Galicia in 1888, Kligler immigrated to the United States in 1901 with his family. After completing his studies in bacteriology and pathology at Columbia University in New York City, Kligler began working as a researcher for the Rockefeller Foundation between the years 1916 and 1920. However, as an ardent young Zionist, he decided to immigrate to Palestine in 1921, where he began managing the laboratories of the Hadassah Hospital in Jerusalem. Shortly afterwards, in 1922, he was appointed as director of the MRU in Haifa, which was formally operated by the British government and partly financed by the American Jewish Joint Distribution Committee.[58]

Kligler was not the only Jewish expert to transfer Western knowledge about malaria to Palestine. Indeed, many prominent Zionist malariologists received their training in leading schools of tropical medicine in Italy, France, the Netherlands, and Britain before arriving in Palestine. However, what distinguished Kligler from other experts was his involvement in several colonial scientific endeavors relating to tropical medicine in Palestine and his role in initiating the collaboration between them. The MRU not only represented a combined British and Zionist effort to fight malaria, it also often conducted surveys for and with the Malaria Survey Section, a private agency funded by the Rockefeller Foundation, which similarly collaborated with the Mandate government. According to Sufian, the joint reports of these institutions were regularly submitted to the British Health Department of Palestine. These reports also supplied important information on indigenous beliefs and practices and offered a demographic, economic, epidemiological, topographical, and climatological mapping of the country from a Western colonial perspective.[59]

During his work in the MRU, Kligler developed several innovative methods to confront malaria in Palestine. Among other things, he introduced several species of fish to the country which were supposed to eradicate the *Anopheles* mosquito, among them the *Gambusia*, which was imported from the United States (and is considered today as one of the hundred most invasive species in the world). Simultaneously, Kligler argued for the low efficacy of quinine and the planting of eucalyptus trees to fight the mosquito, both ideas that were prevalent at the beginning of the century.[60]

Notwithstanding its updated appearance and innovative methods and techniques, tropical medicine preserved many old climatic premises. These included notions regarding the role of climate in generating diseases and longstanding metropolitan approaches toward non-temperate climates. In addition, as we have seen, the implementation of tropical medicine for the purpose of fighting malaria in Palestine was not an exclusively British endeavor. On the contrary, it involved the cooperation and support of Zionist institutions and individual experts who, throughout this process, adopted the British colonial gaze toward the medical conditions of both the country and its indigenous population.

MEDICAL CLIMATOLOGY AND THE FUTURE OF JEWS IN PALESTINE

The historian Warwick Anderson emphasizes how medicine was used in Australia as a discourse of settlement as much as a means of knowing and understanding disease.[61] Likewise, the body of medical knowledge developed among Jewish physicians in Palestine during the first half of the twentieth century mirrored the discomfort of some of the population, in particular their alienation within—and even fear of—their new homeland.

However, while tropical medicine initially developed in the imperial context was generally meant to protect colonizers from the illnesses associated with warm environments, Jewish settlers in Palestine wished not only to protect themselves from the local climate but also to take root in this country and indeed to *acclimatize* to it. As we have seen, the Zionist discourse on the effects of climate on Jewish bodies sometimes contained allusions to the ancient mythology of this people in Palestine. The physician Hilel Yaffe, for instance, cited a famous Talmudic verse, writing that "the air of the Land of Israel creates enhanced wisdom" (אוירא דארץ ישראל מחכים)[62]—"but," he added, "it also heals."[63] The allusion to ancient Jewish texts by medical studies serves as another example of what Novick describes as a fruitful relationship between mystical pasts and utopian, rational futures, which were common among national movements and especially within Zionist thought.[64] The allusion to the Jewish mystical past in Palestine was, of course, also meant to provide "historical" evidence for the "nativeness" of Jews in Palestine.

Nevertheless, as we have already seen, romantic depictions of the climate in Palestine did not always match the reality. This discrepancy was related both to climate itself and to its supposed healing qualities. Isaac Kummer, the protagonist of Shmuel Yosef Agnon's classic novel *Only Yesterday,* said of the Zionists in his hometown, "they'll give you prooftexts from the Talmud that the air of the land of Israel is healing, but when they travel for their health, they go to Carlsbad and other places outside the Land of Israel."[65] Indeed, one of the central medical traditions utilized by Jewish physicians as a source of reference for their climatologically-related practices in Palestine was the work conducted by physicians in Central European natural healing resorts.

Natural healing resorts were common in Europe. However, they also retained strong colonial significance. According to Jennings, hydrotherapy, for instance, had by the first half of the nineteenth century positioned itself as one of the few countermeasures to the alleged degenerative and enervating effects of warm climates. Indeed, the connection was so unmistakable that Jennings writes that the town of Vichy with its natural spas owed much of its growing prosperity and development to colonial expansion. Furthermore, in their remarkable guide to spas for colonialists from 1923, Serge Abbatucci and J. J. Matignon—practitioners of both tropical medicine and hydrotherapy—explained that the Frenchman in the tropics faced three central dangers: the ethnic threat, the pathological threat, and the climate threat. Hydrotherapy spas were supposed to ward off all three, according to the French doctors.[66]

In the Jewish-European context of the late nineteenth century, medical spas initially reflected the desire of this minority to become part of the local professional and cultural elite, while eventually demonstrating their failure to do so. As Mirjam Zadoff demonstrates, healing resorts such as Carlsbad, Marienbad, and Franzensbad in central Europe were extremely popular among the European bourgeoisie. These were places that were identified with relaxation, leisure, and social encounters, and as such they also became important sites of social and cultural mobilization for Central European Jews.[67]

Perhaps more important in this context was the fact that Jews composed 75 percent of the university lecturers whose research and practice focused on hydrotherapy and medical climatology at the University of Vienna. Zadoff explains this by pointing out that, although since 1867 discrimination against Jewish scientists was illegal in the Austro-Hungarian Empire, Jewish medical experts had better chances of developing their careers in small medical specialties where competition was markedly lower. In addition, the fact that medical climatology in spas was considered new and innovative, while at the same time promising a respectable income (largely at private spas), made it particularly attractive for Jewish physicians.[68] The German-Jewish physician Theodor Zlocisti particularly praised the work of medical doctors of Jewish descent who contributed to this field of medicine. In his book *Palestine's Climatology and Pathology* (*Klimotologie und Pathologie Palästina's*), published in 1937, he wrote: "today we talk about the effects of climate on the body in the Alps thanks to the work of—especially—Jewish physicians such as Angelo Mosso, Nathan Zuntz, Adolf Loewy."[69]

As is well known, during the 1920s and 1930s a large group of educated middle-class Jews, who had largely obtained their education in German-speaking universities, settled in Palestine. Many of them were physicians. Nissim Levi and Yael Levi, who collated statistical information on Jewish physicians in Palestine from the late eighteenth century until 1948, argue that during the first half of the twentieth century, the most common medical schools among Jewish physicians in Palestine were those of the universities of Berlin, Vienna, Freiburg, Munich, and Heidelberg.[70]

Unlike the previously discussed romantic views of Palestine's climate and the acclimatization of Jews to it, these physicians often presented a rather distorted view concerning the health implications of the warm climate on Jewish European bodies. I would argue that while in Europe, physicians needed to fight scientific and popular racist views, and this led them to explain their so-called inferior characteristics by stressing the regenerative potential of the climate and the environment of Palestine. Nevertheless, as European settlers in the Middle East who, from 1917 (and, as we have seen, even earlier) were receiving the official support of the British Empire for their settlement project, Zionist physicians and laypeople frequently demonstrated feelings of anxiety and disgust toward their new environment.

In making such a claim, I follow the analysis of Dafna Hirsch, who in her influential work on hygiene in Mandate Palestine stresses the ambivalent position that acculturated Central European Jews held between the Orient and the Occident. According to Hirsch, East and West were used interchangeably by this group as dynamic categories that continuously constructed and reconstructed one another. Explicitly, she argues that during the Mandate period Jews who adhered to Western values and behaviors in Palestine often constructed their own occidentalism by orientalizing other groups (such as Arabs, Sephardi and Eastern European Jews). Nevertheless, while Hirsch discusses these categories in relation to general hygiene education in Palestine and analyzes them as embedded in a modernizing discourse, in my work I position these categories specifically in relation to climate in a settler-colonial discourse.[71]

One of the main issues of complaint and concern among Jewish physicians in Palestine was the *hamsin*. This climatic phenomenon caused dry, hot, and sandy desert winds, which were believed to trigger various physical and mental pathologies. Zlocisti, who dedicated more than twenty pages to describing this phenomenon and its implications in his book, wrote that "it would not be an exaggeration to say that no one feels physically or mentally comfortable in *hamsin* days."[72] He described the mental effects of *hamsin* as including "nervous and mental exhaustion, fatigue, irritation and nervousness. The multi-layered intermediate forms of this phenomenon show themselves in uncertainty, insecurity of memory, loss of eagerness, loss of energy, loss of the capability to concentrate, loss of courage, loss of enthusiasm for work and life."[73] Zlocisti added that this type of weather very often led to explosions of rage and senseless fits, followed by contrition and remorse, or alternatively was characterized by defiance and intolerance in reaction to the most trivial of stimuli. Influenced by contemporary bigoted views, he added that people with mentally unstable characters and women in their forties were especially prone to these reactions.[74]

The increase in complaints concerning the local climate and the environment was most evident from the 1930s, and was partly a result of the consolidation of Jewish medical institutions in the country during these years that, as in other settler-colonial cases, took an ethnic-racial separatist approach towards local populations.[75] As we have already seen, Jewish health institutions often cooperated with the British authorities, but usually remained financially independent of them and relied mainly on funding from Jewish donors abroad. The two main Zionist health organizations in Palestine during the Mandate were the Hadassah Medical Organization, established in 1920 and sponsored by North American Jewry, and the health insurance provider of the Zionist labor union, *Kupat Cholim Clalit*, established in 1911 and supported by the JA and the Jewish National Fund (JNF).[76] Palestinians, on the other hand, relied mainly on British health institutions as well as on a few Christian mission-ary charities. Sufian argues that the resulting situation of this institutional division was "an autonomous, well-organized, and relatively well-financed health system run by the Zionists, and an Arab population dependent upon a limited British health system and missionary efforts." According to her this condition "maintained and perhaps even deepened the social gap between the two communities."[77]

The enhanced negative medical discourse on climate during these years may have also been derived from the recent improvement in the country's sanitary conditions which, as Hirsch argues, allowed the Jewish settler population to focus on "new" concerns.[78] Finally, the escalating Arab-Jewish conflict was most likely a central element that influenced the Jewish sense of climatic discomfort in Palestine. In 1936, this conflict reached a boiling point that was expressed by six continuous months of violence from both sides and a general Arab strike in labor and transpor-tation that shook the local Jewish economy.

Hirsch and Sufian demonstrate in this context how general ideas and measures that were meant to protect the health of the Jewish community in Palestine frequently contrasted its lifestyle with that of the local Palestinians.[79] Instead, I argue that the medical discourse that linked cli-mate with the culture and appearance of the local Arab population was, in fact, more complex. While physicians indeed sometimes contrasted the lifestyle of Jews with that of Arabs when it came to questions of climate, they more often ignored this population altogether. This was perhaps a result of the belief, expressed earlier by Warburg, that Arabs did not need to go through any kind of acclimatization processes in their "natural" envi-ronment. As we shall see below, besides contrasting and ignoring them, at other times Jewish physicians actually aimed to study local Arab tradi-tions and habits on how to cope with Palestine's climate.

The approach of Jewish settlers to the local climate often also depended on the intensity of their ideology. Thus, statements reflecting the aliena-tion of settlers from the local environment were especially vivid in the 1940s when many Jewish refugees were forced to go to Palestine without ideological motivation. For example, Margaret Bergel, a German-Jewish refugee who wished to return to Europe after the War, wrote in an applica-tion to the United Nations Relief and Rehabilitation Administration:

בנך לוהט השמש ?

Figure 3: Illustration of the summer heat in the daily
Ha-Mishmar newspaper from 1947. Courtesy of Yad
Yaari Research and Documentation Center, Al
Ha-Mishmar Collection.

In 1933 I immigrated to Palestine from Germany; already in 1935, it
appeared that my health suffered seriously from the local climate, and therefore I returned for a few years to Germany in 1936 where I felt perfectly
healthy from the very first day. Of course, for the well-known political reasons, I could not stay there, and returned to Palestine having no other possibilities. Very soon, I fell sick again but never had the means of going abroad
on a trip again. The war broke out, every summer here meant a crisis to my
health, and all doctors whom I consulted agreed that I . . . cannot stand the
climate of Palestine, and ought to go back to Europe as quickly as possible,
'immediately after the war', they said.[80]

Another refugee, Gershon Mankowitz, who wanted to leave Palestine for
the USA wrote in 1949: "Does no one care about my fate? The climate in
this country is not suitable for my health. My doctor advised that I must
leave the country to be cured. I have a sister and many relatives in the
United States. They can help me."[81]

As previously mentioned, one of the formal objectives for the utilization of medical climatology in Palestine was to determine how and whether Jewish colonization would be possible. Already in 1912, the physician Aaron Sandler had asked in this context: "Granted that our people can develop politically, socially and economically in Palestine, can they also develop physically?"[82] In 1939, when optimism about the Jewish ability to acclimatize in Palestine was fading, the physician R. Kazenelson wrote that for the Jewish settler to become properly rooted in Palestine, he would need to adjust to the new climate and the new environment, and accordingly develop a new lifestyle. To clarify the medical aspect in his point, Kazenelson added that "when referring to adjustment, we clearly mean . . . how to prevent diseases."[83] Likewise Joseph Davidson, the medical inspector of the JA's immigration department, explained that whereas most people are immune to diseases in their native environments, this "natural immunity" is confounded when they move to new environments. In such cases, Davidson suggested, "one should try to maintain their health, [and] be cautious of diseases and exhaustion, especially if [in the process of immigration] changing to a physical occupation."[84]

Knowledge of the effects of the natural environment and the potential diseases it might harbor, coupled with an awareness of, and even personal responsibility for, maintaining a healthy lifestyle, were understood to be essential to the success of Jewish acclimatization to Palestine. The physician Jacob Seide wrote that unlike other organisms, humans have the wisdom to adjust themselves to external conditions in order to reach a harmonious relationship between their biological characteristics and natural environment. To achieve this goal, he argued, "humans need to acquire a deep knowledge of all the elements that could influence their lifestyle, their health, and their cultural standards."[85]

As a result of this approach, many medical publications during the 1930s and 1940s were co-edited by physicians and meteorologists and addressed a popular audience, thereby intending to educate the masses into maintaining a better state of health in their new environment. These books included, for instance, titles such as *Health of the Nation* [*Briut Ha-Am*], *Health of the Immigrant* [*Briut Ha-Oleh*], and *Public Health* [*Briut Ha-Zibur*]). Analyses of these texts demonstrate that the penetration of hygiene to everyday life was usually discussed in two central contexts: the sphere of labor and the domestic sphere.[86]

With regard to labor, Seide reported on foreign experiments, asserting that the most productive temperature for physical work was 16 degrees Celsius, and for mental work, 4 degrees Celsius. However, accepting these conclusions would imply that the climate in Palestine was unsuitable for physical work for more than six months of the year, and moreover deem it categorically unsuitable for mental work. Seide thus had to settle on the results of another experiment, which argued that the best temperature for such activities was not above 24 degrees Celsius.[87]

Likewise, Zlocisti argued that July and August were the hardest months for laborers in the Jordan Valley. Typical of his dismissive approach to women, he added that this was three times the case for women, "even though," he claimed, "their work is much less demanding."[88] The physician Fritz Kahn similarly expressed his concern towards the ability of pupils to study during the warm summer months. Kahn asked "where, in what country, at the heat of 28–30 degrees [Celsius], are children expected to complete their homework after spending six hours at school?"[89]

The clothes of the Zionist laborer were also examined in relation to the climate in Palestine. In an article on the topic, Professor Walter Strauss, director of Hadassah Hospital in Jerusalem, complained that the human desire to look attractive had surpassed basic climatic needs. Therefore, he proposed a few basic principles to ensure that settlers' outfits would also fulfil their climatic needs. Strauss argued particularly for the wearing of light colors that were capable of reflecting sunlight, although he ultimately claimed ventilation to be of utmost importance. He advised adopting some traditional Palestinian costumes, while in the same breath recommending a unique Russian-style shirt, which he thought would complement local conditions. This shirt was to be worn outside the trousers. Strauss stressed that it should also be worn without a belt to allow as much ventilation as possible. In addition to what he called "horizontal ventilation," as enabled by the neck, armpit, and stomach area, there would also be "vertical ventilation," since the shirt was designed to have small holes at the back.[90]

The majority of guidelines for the acclimatization of Jewish settlers to Palestine, however, were focused on the domestic sphere. For instance, in May 1936 the newspaper *Davar* published the "Ten Commandments for

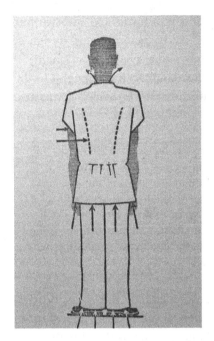

Figure 4: Drawing of shirt appropriate for Palestinian climatic conditions as suggested by Walter Strauss, c. 1930s. With courtesy of the Strauss family and Olympia Press.

Hot Summer Days." The recommendations by Dr. M. Shechter called upon readers to avoid direct contact with the climate as much as possible and to ensure maximal separation between the body and the surrounding air. Shechter recommended shutting all windows, shutters, and doors in living spaces and, furthermore, covering the body with woolen clothes "that absorb perspiration."[91]

Hygienic concerns, mirroring domestic recommendations, also embraced the Jewish cuisine in Palestine and were most often addressed to housewives. For instance, a cookbook from 1937, entitled *How to Cook in Palestine?* and published by the Women's International Zionist Organization, advised readers to abandon their traditional recipes in favor of culinary acclimatization. This book was written in Hebrew, English, and German, reflecting the background of both the author's and her target audience. Dr. Erna Meyer was a German-Jewish economist who specialized in household economy. During the interwar period she achieved a successful career in Germany, which included the publication of another book entitled *The New Household: A Guide to Economical Housekeeping,* which saw forty-two German editions and was translated into various other languages. Meyer was also involved in the planning and design of modern kitchens, based on the premise of rationality and functionality.[92] This notion was, as we shall see in the next chapter, fundamental to the ideas of the modern movement in architecture. Nevertheless, it was also gaining attention in relation to the consumption of food.

> What shall I cook? This problem, the concern of housewives the world over, is particularly acute in our country. The differences in the climate and necessary adjustments arising out of these differences compel the European housewife [in Palestine] to make many drastic changes in her cooking—a change not so easy to achieve as it would seem. . . . [In] most families the adjustment is slow, unwillingly and incomplete. . . . Some of these [old] habits are not only injurious to the health of the family, but in addition burden the housewife with unnecessary work.[93]

Among other advice, Meyer suggested replacing "European fat-rich food," which, according to her, included beef and butter, with local fruits and vegetables. As she explained, "cooking suitable to the climate must place vegetables, salads and fruits in the foreground."[94] Seide and other experts similarly suggested substituting the nutritional value of fats with tropical fruits, such as the avocado and guava, which they described as good sources of protein, carbohydrate, vitamins, and minerals.[95]

Thus, as chapter 4 discusses, the physical acclimatization of Jewish European settlers in Palestine required the consumption of acclimatized foods, a symbiosis that reflected the foreign characteristics of the Zionist project in Palestine. The promotion of acclimatized fruits by experts reflected not only the new and foreign technologies promoted by the Yishuv (such as the use of irrigation, machines, and fertilizers) but also the integration of other modern Western fields of expertise within general medicine, such as hygiene and the science of nutrition.

Nutritional science was developed in the second half of the nineteenth century in Europe and in North America. One of the foundational premises of this new science was that there was a strong link between diet and health. Following the initial development of this field, dietary intervention by the state became a common phenomenon. For instance, during times of war, contemporary principles of nutrition were applied to both soldiers and the civilian population. In peacetime, it was manifested in the provision of school meals, hospital diets, and the education of housewives on nutritional issues through the press, cookbooks, hygiene guidebooks, and other media.[96]

According to Yael Raviv, by adopting this scientific approach Jewish experts were able to present food as "productive," "rational," and "useful."[97] Dr. Shechter, author of the "Ten Commandments for Hot Summer Days," wrote elsewhere about how food should be consumed in Palestine's climate. He argued, for instance, that eating a small breakfast, a large lunch, and a medium dinner, as common in many European countries, would cause more harm than good to the Jewish laborer in Palestine. Instead, he suggested a protein and calorie-rich breakfast, which included dairy products and eggs, and a dinner containing meat, which he claimed to be easier to digest in the post-work hours. His advice about meat was supported by the biblical quote "And Moses said, 'When the Lord gives you in the evening meat to eat.'"[98]

However, as also shown by Efrat Gilad, when having to decide on a different type of nutrition in the new homeland, some experts recommended the adoption of local eating habits.[99] A *Davar* article from 1925 entitled "The Science of Nutrition" explained: "Nutrition has been grounded in a scientific base only in recent years. . . . [F]or us [in Palestine] this science is particularly important, since we are moving towards new living conditions and a new climate. We must look to the local inhabitants and their foods, which are the result of instinctive selection over generations."[100] Mordechai Brachiyahu, the director of the Hadassah Medical Organization Department, similarly wrote, "what is good and natural to the land's climate, so far as it is part of the Arab cuisine, deserves our attention."[101] But, he added, "it must be corrected in a civilized manner,"[102] thus reminding his readers that becoming native in Palestine did not mean abandoning Occidental standards.

Besides food, the amount of water one needs to drink, and the amount of sweat and urine produced in the warm weather, were also measured and discussed in hygiene guidebooks for the public. Seide saw no reason to limit the amount of water for a healthy child or adult during the summer days. This was because the body is able to dispose of excess water but cannot adjust to a lack of water for too long. However, to prevent any misunderstanding he specified the exact minimum amount of water (one and a half liters) that a person should drink during the Palestinian summer. The meteorologist Rudolf Feige, moreover, suggested the addition of cooking salt into the water on hot days, since excessive sweat could leach vital minerals such as sodium, calcium, and potassium.[103]

Similar guidelines on the "right way" to eat, drink, dress, work and rest in warm climates were common in popular medical publications in the British colonies, and it is likely that this literature served as a role model for Jewish physicians' medical guidebooks in Palestine. Books such as *Health in Africa; a Medical Handbook for European Travellers and Residents* (1897) and *Health and Sickness in the Tropics: A Guide for Travellers and Residents in Remote Districts* (1913), among other publications, discussed the effects of warm climates on the temperature, respiration, urine, nervous system, and menstrual cycle of Anglo-Saxon bodies.[104]

However, such publications were not only similar to those produced by Jewish physicians in Palestine. In some cases, Jewish physicians also alluded directly to contemporary medical climatological research conducted mainly in settler societies, such as the United States and Australia. These countries were presented as successful examples for the acclimatization of European people to warm climates. Seide, for instance, wrote in encouraged surprise that despite the tropical conditions in Queensland, white settlers had not yet degenerated there.[105] Moreover, Zlocisti's bibliography referred to scientific publications of his time such as, for example, *The Metabolism of White Races living in the Tropics* and "Tropical Australia and its Settlements."[106]

Expressing fear, disgust and complaint about the hazards of the local climate stressed the foreign and European origins of Jews in the Palestinian natural and cultural environment, and moreover helped experts and laypeople to position themselves as inherently superior to the local population. Nevertheless, as we have seen, this was not the only approach Zionist physicians took toward the local climate in Palestine. When facing difficulties in adjusting to the local climate, Zionist physicians tended to emphasize the need of settlers to become rooted in the country and thus, in their studies, focused on practical solutions that were supposed to help them achieve this goal. These solutions at times even included a romanticization of the local Palestinian lifestyle. More than reflecting an objective physical reality in Palestine, medical climatology and its corresponding popular discourses mirrored how physicians and laypeople perceived the Jewish European body in a changing physical as well as social, cultural, and political reality.

In this chapter, I have focused on Zionist medical perceptions of the climate in Palestine during the first half of the twentieth century. Throughout this period, medical views on the climatic conditions in Palestine were not always consistent and tended to reflect the shift in Zionist perceptions towards both the Jewish body, as well as the new Jewish homeland and its Arab inhabitants. In all three sections of this chapter, as well as in the medical traditions and notions they represent— race science, tropical medicine, and medical climatology—we encounter the role of medicine and climate in reflecting the liminal position of European Jews as a minority aspiring to liberate itself from a long history of persecution and oppression, on the one hand, and the tendency to embrace an amalgamation of colonial ideas and practices as part of the Zionist attempt to obtain a non-European territory, on the other.

3. Warm Palestinian Climate— Cool Jewish Spaces

As discussed in the previous chapter, climate in Palestine was sometimes understood as a hazard to health that could, if not properly addressed, pose a potential threat to public well-being. As a result, Jewish physicians aimed to educate the masses into adjusting their everyday lives to local environmental conditions. This medical discourse focused mainly on clothes, nutrition, and working arrangements. However, the fear of local climate and its implications for the body was so central to the Jewish settler experience in Palestine that it transcended the medical discourse and became an integral part of other professional discussions, mostly in the fields of technology in general and of architecture and planning in particular.[1]

The organization of the built environment, as well as of its surrounding landscapes, in accordance with the local climatic conditions in Palestine played a crucial role in the Zionist project throughout the period addressed in this book and beyond. As I demonstrate in this chapter, attempts to counter and alleviate the perceived enervating effects of the local climate in the built environment were influenced by new technological innovations. For instance, Zionist settlers aspired to apply groundbreaking (and at times even utopian) refrigeration technologies, developed in other colonies located in warm regions and intended to allow settlers to continue to pursue their European lifestyle. On the other hand, Jewish architects were also interested in exploring Palestinian methods of coping with the local climate. Thus, many contemporary professional discussions focused on the suitability of artisan building methods to the climate in Palestine. Nevertheless, despite their desire to appropriate local building motifs while sometimes presenting them as ancient biblical ones, Jewish architects eventually tended to rely on modern Western techniques, even when these techniques proved less efficient in the local climate. As outlined in the introduction, both the wish to absorb local practices and the tendency to reject them were common settler-colonial approaches.

"THE REFRIGERATING INDUSTRIES"

Mike Hulme writes that during the twentieth century, climate was often referred to as a force that needed to be conquered. This aspiration was mostly achieved by improvements in medicine and the refrigerating and air-conditioning technologies that aimed to remove some of the direct physical fears that tropical and other warm climates presented to non-indigenous populations.[2]

This modern technology, which made artificial indoor environments independent of the weather outside, developed in warm climates inhabited by white settlers from the mid-nineteenth century onward. Historians agree that there were two main reasons for the development of this technology. The first was the general aim to support Europeans in warm climates while improving their sense of comfort and ensuring their productivity. The second was related to the desire to preserve food (especially meat) in warm climates, and to transport it from the colonies to Europe.[3]

In addressing the first reason, Rebecca Woods points out that for the British,

the technology of artificial cold and manufactured ice offered to solve the problem of tropical climates. Standardizing the temperature of the empire was one of the most exciting potential applications of [this] novel technology. . . .. For enthusiasts of artificial refrigeration, a reliable supply of cold and ice was all that was needed to make India and other tropical colonies habitable and risk free for British colonists.[4]

Indeed, the first ice maker is believed to have been developed in Australia in 1855. Furthermore, at the time of its appearance, it was said that: "It is in hot climates ... that the full value of the invention will be felt. Ice within the tropics will soon be looked upon as a necessary of life, as much so at least as fuel is a necessary in the winter of temperate regions."[5]

As we have already seen in other aspects, Jewish settlers in Palestine sometimes used a similar rhetoric to that of other Europeans in warm climates. In 1902, Herzl had likewise suggested utilizing air conditioning in his novel *Altneuland* when one of the characters, the architect Steineck, discussed a utopian machinery employed by Jewish settlers in Palestine to confront the local climate.

> This is a warm country. From here down along the Jordan ... the country is pretty well heated the year round. We have therefore developed the refrigerating industries. Understand? On the principle that the best stoves are to be found in cold countries. ... We, for our part, have provided ourselves with plenty of ice for the warm weather. In the heat of the summer, for instance, you will find blocks of ice even in the most modest homes. ... [T]hrough science we have learned how to make ourselves more comfortable and healthier. Understand?[6]

Herzl's reference to the "refrigerating industries" indicated, on the one hand, his familiarity with the most advanced technology of his time. On the other hand, the allusion to this technology should also be discussed via its fantastic connotations. As Gail Cooper points out, during the first decades of the twentieth century, artificial climates were frequently mentioned in utopian novels and other fictional accounts of advanced societies.[7] Indeed, Herzl's novel did not offer a practical plan for founding the imagined Jewish state. Instead, he wished to inspire his readers with the possibility of a Jewish homeland by emphasizing the liberating power of science, medicine, and technology, which, according to him, would transform Palestine into a fertile land, and similarly, transform the Jews into a people like those of all other nations. The utopian state as it was presented in *Altneuland* absorbed the accomplishments and achievements of modern Western nations while alluding to a colonial rhetoric that implied that where "nothing" exists, modern innovations can be implemented with great success. Even more important for our subject matter was the fact that Herzl's allusion to the refrigerating industry referred to the nineteenth-century orientalist and colonial discourses of environmental anxieties and suggested that, in his eyes, Jews had more in common with European colonialists encountering an unfamiliar environment than with a people "returning" to their historical homeland.

In addition, as Barak reminds us, another important advantage of cooling technology was its potential ability to facilitate greater productivity among white people in warm climates. Barak quotes in this context an anonymous European who wrote in the 1860s that by using such a technology, "the miseries that press upon the whiteman [sic] in a hot climate might be alleviated, and we might then really see what northern muscles can affect in southern regions."[8]

One of the earliest visionaries of air conditioning was Florida physician John Gorrie, who as early as 1842 wrote that air conditioners were needed "to counteract the evils of high temperature and improve the condition of our cities."[9] Gorrie was concerned with the health of Americans in the southern United States, but he also appears to have been bothered by their productivity levels. In 1854, he wrote that "the cheap and abundant productions of artificial cold would enable many industrial arts to be carried on advantageously in warm climates."[10] The role of air conditioning in maintaining and enhancing the productivity of Westerners in warm climates was specifically relevant to settler societies, since the use of local labor was supposed to be avoided in these places; instead, settlers usually wished to create a self-sufficient economy.

This logic was also true for the Yishuv. As mentioned, one of the main concerns for physicians and patients in Palestine was the *hamsin*. In the 1930s the German-Jewish physician Fritz Kahn lamented the flawed productivity of Jewish settlers in such weather, and advocated building underground shelters "against the air" while simultaneously installing "air conditioning"—a technology that he believed would be of particular value to hospitals and children's homes.[11]

When writing about the negative effects of the *hamsin* on health and productivity, Kahn also suggested that the meteorological services should warn the population of this climatic phenomenon in advance through the use of special alarms. Kahn complained that not only did such a

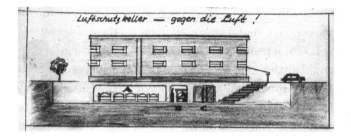

Figure 5: Fritz Kahn, illustration of air shelters 'against the air!,' 1936. Courtesy of the Leo Baeck Institute Archive at the Center for Jewish History, New York.

technology not yet exist in Palestine but, on the contrary, instead of avoiding activity on *ḥamsin* days, some schools maintained their regular curriculum, endangering the health of the children. Moreover, while the so-called reduced productivity of local Arabs in warm climates was often critically presented by Westerners as reflecting their alleged passivity and indolence, Jewish settlers were advised (and even warned) by medical experts not to be physically active in such weather in order to protect their health.

As we have seen, the desire for an air-conditioning technology that would alleviate the heat and make life and work more convenient for Jewish settler society in Palestine initially appeared in 1902 and prevailed throughout the century. Nevertheless, the actual introduction of this technology into the country came relatively late. The first factories to produce industrial air conditioners for public spaces such as hospitals, hotels, cinemas, theaters, and factories were only established in 1955.[12] However, by 1956 they had already begun to produce air conditioning for domestic use (first for houses in the kibbutzim of the Negev), as well as to export their products to Cyprus, Brazil, French Guiana, Senegal, and Kenya, among other countries. These air conditioning factories were sponsored by foreign investors from other settler societies, such as South Africa and the United States.[13]

As mentioned earlier, this technology also made more food available in wider geographical dimensions. One of the first practical applications of industrial refrigerators was Britain's meat industry. During the late nineteenth century, the growing population in Britain surpassed the domestic sources of available meat. Simultaneously, Australia was over-flowing with cattle and sheep. Newly invented engines producing ice solved this problem by enabling the transportation of meat across oceans from the 1870s onwards.[14]

A newspaper article about the first Hebrew ice factory published in 1928 declared that "ice rooms" were a blessing for the food trade as well as for consumers, as they enabled people to obtain fresh and seasonal food at any time and assisted tremendously in the development of the dairy industry.[15] While the development of advanced air conditioning technology in the Yishuv mostly exceeds the time frame of this study, ice technology and refrigerators for the purpose of cooling food were, in fact, already available during the first half of the twentieth century. Ice machines were generally introduced into large urban centers in the Ottoman Empire starting from the last third of the nineteenth century, and could be found in places such as Jaffa, Haifa, and Jerusalem in subsequent decades.[16] In fact, the first Jewish ice factory was opened as a result of a lack in ice following the Arab-Jewish riots of 1921, when Arabs refused to sell their ice to Jewish hospitals.[17]

In private Jewish households in Palestine, ice machines (or iceboxes, as they were often called) began appearing in the 1930s. This early form of refrigerator was used for the daily preservation of fresh meat and dairy products, and since it was not powered by electricity, it required a continuous supply of large ice cubes. As Efrat Gilad reminds us, Jewish European settlers preferred consuming meat and dairy products that were not abundant in the local environment. In addition, as we saw in the previous chapter, these foods were not only seen by experts as difficult to digest in the local climate but were also difficult to preserve in it. Gilad quotes Erna Meyer and other nutrition and health experts pointing out the dangers caused by foods that had rotted during Palestine's summer. One expert wrote, for instance, that "during the hot and humid weather one is bound to hear of a certain number of cases of food poisoning due to spoiled meat."[18]

The demand for ice seems to have increased in tandem with the Jewish population in Palestine. Moreover, as mentioned above, during the 1940s a growing number of Jewish refugees arrived in Palestine without much

Zionist ideological motivation. During this period ice consumption increased, as it was seen as absolutely essential for the maintenance of the settlers' Western lifestyle and cuisine. An article in the Hebrew press from 1946 reported that, according to a recent census, 40,000 residents in Tel Aviv were consuming ice on a daily basis, while 4,200 of the city's wealthier residents possessed their own mechanical refrigerators. In addition, it was reported that Tel Aviv had seven ice factories and forty-four delivery men selling ice from carts on the street every day. During 1945, the total consumption of ice was 210 tons per day, which in 1946 rose to 300 tons per day; this number, according to the article, was expected to rise by 50 per cent in the following year.[19]

I have been unable to find historical evidence for the use of iceboxes and refrigerators for domestic use in Arab households in Palestine. However, this does not mean that they did not exist. Recent studies focused on Palestinian capital and especially the local Arab bourgeoisie during the Mandate period tell the forgotten story of this socio-economic group that advocated for Palestinian nationalism via economic growth and Western science and technology. In this context, Sherene Seikaly mentions Palestinian businessmen who owned, among other enterprises, local ice factories.[20]

Since European Jews in Palestine often collectively insisted on their old foodways, during these years the Hebrew press repeatedly reported on the massive damage to food and the subsequent disappointment and outrage expressed by Jewish settlers when problems occurred in ice delivery on summer days (complaints before weekends and holidays were even more common). One columnist fumed: "Will [the municipality] finally put an end to this scandal? Why do the municipality and the authorities allow such an abuse of the residents?"[21]

When an ice factory underwent refurbishment in July 1945 and failed to deliver the usual amounts of ice to Tel Aviv, masses of residents were reported outside of the factory building, demanding the ice they had been promised. The large queue that formed outside the factory soon descended into a fistfight that was defused only following police intervention. Furthermore, the article stated, thousands of people's food was spoiled.[22]

Of course, refrigerating food also had broader economic advantages for the market. A report of the agricultural experimental station in Rehovot from 1945 stated that, thanks to the use of refrigerators, various fruits could be sold beyond their ripening season. For instance, citrus fruits that ripened in Palestine during the rainy season could be kept in refrigerators at 10 degrees Celsius from March until July and be sold in the summer.[23] Cooling houses were also useful when the crop yield was bountiful, and the number of fruits too large to be sold for fear of flooding the market. A Fruit Growers' Union handbook from September 1942 stated, for instance, that during the summer months, 50 tons of peaches, 50 tons of pears, and 330 tons of apples were stored in industrial refrigerators.[24]

The amount of water required for the creation of increasing quantities of ice must also be considered, bearing in mind the scarcity of water in Palestine, which will be discussed in the following chapter. As we have seen here, Jewish European settlers' fear about the local climate manifested not only in contemporary medical discourses but also in technological aspirations to alleviate the indoor heat during summer and to prevent it from spoiling food, especially at times of *hamsin*. Further technological solutions for managing the local climate's impact on Jewish European bodies were assumed by architects and planners.

THE GARDEN CITIES OF YESTERDAY

Prior to the full development of air conditioning as a commonly used mechanical solution to confront the local climate, Jewish experts in Palestine tended to place great emphasis on planning houses and streets in accordance with the prevailing winds and sun direction. In 1909, a few dozen Jews from Jaffa decided to build Ahuzat Bayit, a new Jewish residential neighborhood outside the old city walls that would later become the city of Tel Aviv. While the first houses and streets were being built, the founders of Ahuzat Bayit agreed that they would need a town plan for their growing neighborhood. The first plan for the neighborhood was created by engineer Avraham Goldmann. Later Arthur Ruppin, the director of the Palestine Office of the ZO, composed the first town plan regulations.[25]

Since Ruppin had very little experience and knowledge of planning, he turned to the architect Oskar Marmorek, who was also a close friend of Herzl, for advice on professional literature on urban planning. Marmorek's suggested list included Ebenezer Howard's seminal book *Garden Cities of To-morrow* and Josef Stübben's *Handbuch der Architektur: Der Stadtbau* (Architectural Handbook: City Planning).[26] New ideas on city planning, such as those expressed in *Garden Cities of To-morrow*, developed in Europe as a response to the Industrial Revolution and the growing demand to improve the polluted and crowded neighborhoods of the proletariat. In Palestine (as well as in other colonies located in arid regions), the notion of the garden city received renewed orientalist enthusiasm, and mainly served to transform these dry ecosystems into so-called lush and green ones.[27]

Scholars identify the copying of European spatial models and their import into non-European geographies as a key characteristic of the design of colonial spaces.[28] Moreover, studies of colonial planning argue that Western knowledge was often considered to be more reliable, ensuring "safer," "better" construction, making it the preferred mode for building the houses and neighborhoods of European settlers. Thus, another planning feature identified with colonialism was its tendency to celebrate the "new" and "imported" as opposed to the "old" and "local." As we shall see in the following sections of this chapter, this preference was evident not just in architectural and planning styles, but also in technological and engineering innovations, as well as the utilization of new building materials. A final characteristic of colonial planning involved the spatial separation of colonists and colonized.[29]

In Palestine, this separation was often also reflected in the perpetuated perception of local urban spaces as dirty, congested, and diseased. The historical sources reflecting this perspective are infamously numerous. One example is a letter from 1935 sent by a German-Jewish settler to his family abroad where he described his impression of Jaffa as filthy and crowded, with dilapidated buildings. Nevertheless, he noted that "the closer one gets to the center of Tel Aviv, the more the impression becomes European."[30] In what follows, I demonstrate the ways in which the above-mentioned colonial attributes of the built environment were also intertwined with cultural and medical notions about the local climate in Palestine.

Among its various clauses, Ruppin's regulations for Ahuzat Bayit permitted the construction of a single house on each parcel of land and restricted the attaching of two or more houses together. The minimum distance between houses was to be one meter or more. In addition, only one-third of each parcel could be used for the house itself. The rest of the land was to be preserved for a mandatory garden. The purpose of the regulations was not only to ensure green spaces for the neighborhood, but also to allow the flow of fresh air between its houses.[31] This particular demand was often associated with hygiene concerns, as well as the Zionist wish to distinguish its settlements from local Arab examples. In this context, Hirsch quotes Dr. Asher Goldstein, the author of hygiene training manuals and editor of the health section in the daily newspaper *Ha-Aretz*, who in 1935 wrote:

> Which one of us doctors is unable to produce evidence to the extent of the people's ignorance regarding questions relating to local climate and new living conditions? And, should anyone doubt this, the proof lies in the filth, uncleanliness, and lack of trees and plants. And let none of us settle this issue by comparing ourselves to our Arab neighbors, since we do not live off them, as indeed we have come here to bring the West, the best and most refined it has to offer, not only for ourselves, but also for the entire backward Orient, which must urgently awaken to a hygienic, clean way of living.[32]

The fear of assimilating to the local environment and culture and the perception that the lack of cleanliness and trees were the result of the country's cultural and environmental conditions are evident in many texts written by Jewish experts at the time. According to Goldstein, the local Arabs and the climatic conditions were two sides of the same malignant coin— each side causing the other. His text also betrays the assumption that a huge cultural gap existed between Arabs and Jews, which was to be narrowed by the latter's capacity (and "benevolent" obligation) to civilize the indigenous population. The historian Mark Levine has argued that, like other colonial urban spaces, Tel Aviv was nourished by the discourse of development, and hence it justified its separation from its mother city Jaffa by representing itself as a negation of the latter's existing conditions.[33]

Moreover, greening urban spaces, and especially the practice of planting trees, were widely associated in the scientific literature of the time with the improvement of both health and the climate. Public health reformers in the West often argued that trees could "save lives." For example, the American physician Stephan Smith, who in 1911 became the president of New York City's Tree Planting Association, argued during the 1870s for a correlation between diarrheal diseases and hot summer days. To reduce the temperatures in the city and achieve an immediate "cooling effect," Smith and his fellows recommended planting street trees. Such notions about trees were often also linked with the ambition to fight malaria, as some believed that trees absorbed poisonous gases.[34]

In 1921, following the commencement of the British Mandate in Palestine, Tel Aviv received local council status, and then city status in 1934. At the time, it was the only urban settlement inhabited and exclusively run by Jews in Palestine. In fact, although the municipality was subject to the British town planning ordinances, the actual level of British involvement in the construction of the city was very limited. The waves of Jewish immigrants in the following decades, especially of urban middle class Eastern and Central Europeans during the interwar period, naturally also created a growing demand for houses in this city. As a result of the city's rapid growth some of its leaders were afraid it would develop unsuitably. Therefore, in 1925 the Tel Aviv municipality contacted the Scottish city planner Patrick Geddes, asking him to plan the future districts of Tel Aviv in an area north of the existing town spanning 860 acres.

Geddes was among many experts who had studied in Britain and developed their professional careers in the empire's colonies. In 1914, only five years after city planning became compulsory in Britain, Geddes set out for his first job as a city planner in India. During his career as a city planner for the British Empire, he composed more than fifty reports for various cities in India, Sri Lanka, Yemen, Palestine, and other countries. Geddes believed that the planning of a city must begin with a thorough familiarity with its health standards and environmental, social, and economic characteristics.[35] Yet historian Noah Hysler-Rubin claims that, despite the importance of the survey in Geddes's work, what guided his practice above all was a unique urban theory that he developed in his early years as a planner in Edinburgh, and that he reproduced in various cities around the world. According to Hysler-Rubin, the adjustments Geddes made in the plans of the different places he visited were, in fact, so insignificant that his reports were more similar than they were different.[36]

Geddes's plan for Tel Aviv was based on a grid model, which created a north-south axis crossed by streets running eastward from the sea.

According to this plan, most of the urban traffic was to flow between the north and south so that most buildings would be oriented east-west. This arrangement of streets and houses guaranteed maximal air ventilation, as the winds in Tel Aviv blew from the west during the day and from the east at night. The emphasis on main roads between north and south also minimized direct sunlight on house façades from the south. Hence, this plan created a hierarchy within the streets, ensuring a distinction between "main-ways" and "home-ways." The latter were supposed to be "as few, short and narrow as possible"[37] and to evade constant direct sunlight. Like Ruppin a decade earlier, Geddes also recommended the construction of freestanding buildings to ensure the ventilation of air between houses.[38]

Considering the troubles architects faced in arranging the parcels according to air and light, Geddes's plan does not seem to have aided much in solving these problems. The huge increase in the Jewish population during the interwar period led landowners to sell smaller and denser parcels for higher prices. Ruppin's and Geddes's regulations were creatively interpreted in accordance with the contemporary circumstances and, in general, were not always enforced. This situation resulted in criticism from planners, architects, and inhabitants alike. Some architects claimed that the space between the houses was not sufficient to enable the beneficial ventilation of air. Others complained that the regulations forced planners to waste expensive land that could have been used for constructing more houses. In an article from 1936, architect Dov Karmi wrote: "In the parcellation of our city . . . the purpose of land division was not considered thoroughly, that is, to allow the architect to build a healthy house—a good house. . . . Today we see parcels on which it is impossible to build a good house."[39]

A similar view was expressed by the city's dwellers. In a letter to the editorial board of *Ha-Aretz* from 1935, an anonymous writer criticized the architects' disregard for the regulations, blaming them for the heat and lack of ventilation in Tel Aviv's flats:

> Twenty-seven years ago, when Tel Aviv was founded, and its creators aimed for a small garden city, a suburbia of villas next to Jaffa, they regulated the distance between houses. . . . This is not the case anymore: houses are large because they [architects] are using every inch of parcel; they [the houses] are of three stories and thus we get a surface of walls, and this surface—

besides when the walls turn to the north—heats due to the eastern, southern and western sun, and later the walls . . . project this heat onto the opposite walls, i.e., on the windows of the houses that are open during the summer. This creates a reciprocal heat projection and . . . increases the lack of air in the rooms. Fresh air and wind? They are very limited and sometimes do not even exist.[40]

Indeed, architects who wished to design convenient houses in accordance with planning regulations were forced to come up with creative solutions. In 1945, Karmi suggested three original solutions for the arrangement of houses on a parcel. He called the first solution "the chess plan." According to this plan, houses were built at a proximity of less than two meters, but their façades were not placed on a straight line and did not face each other. According to Karmi, this kind of arrangement created a sense of depth in space and, moreover, left room for a garden.

The second solution was entitled "pair of houses." According to this plan, every two houses were to be attached to each other, thereby allowing a larger space of four meters between each pair. Unlike the first solution, this method recommended placing the houses in a straight line in order to

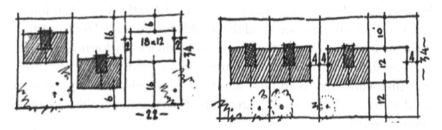

Figure 6: Dov Karmi, "The Chess Plan" (a) and "Pair of Houses" (b), 1945. Courtesy of Ada Karmi and the *Journal of the Association of Engineers and Architects*.

Figure 7: Dov Karmi, "The Continuous Building," 1945. Courtesy of Ada Karmi and the *Journal of the Association of Engineers and Architects*.

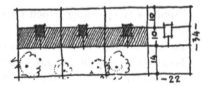

create a larger garden around them. The third solution (and the most preferable, in Karmi's view) was called the "continuous building." According to this method, all the houses were built in a line, and the gardens in front of the houses consolidated into one green strip of seven to eight meters in length. This plan also enabled a large backyard ten meters wide, and the distance to the next row of houses on the same block reached almost twenty meters.[41]

Despite the planning regulations, which were meant to enable maximum ventilation as well as a green environment, Tel Aviv did not develop into a true garden city. Its dependence on private investment did not leave much room for public spaces, and instead the trees and flowers that were planted in private gardens became the main expression of the garden-city dream.

During the Mandate period, and even earlier, most Jews settled in Palestine's towns and cities. Nevertheless, Labor Zionist ideology continued to glorify the rural life in cooperative agricultural settlements. In this respect, the spatial typology of the garden city in Jewish settlements in Palestine was not limited to urban environments. It was also integrated, for instance, into the planned landscape of the communal agricultural settlements of the kibbutzim from the 1920s. Gardens were an integral part of the kibbutz plan. This constructed landscape was uniquely termed *noi*, meaning "gardenscape." The architect Samuel Bikels, a kibbutz member himself, theorized this type of landscape as a new spatial model that connected aesthetic concerns with social, political, and hygienic ones.[42]

Industrial and agricultural zones in kibbutzim were usually separated from residential zones by a large green belt, and some architects asserted that the green belt should be at least sixty meters wide. This area was supposed to be located at the center of the kibbutz, where the dining hall and culture house would also be located. The German Jewish architect Richard Kaufmann, a prominent advocate of garden-city ideas, argued that the wind direction should be studied in each settlement and that living zones should face the direction of the wind so that cold breezes would enter before moving onwards to the industrial and agricultural zones. In this way, air hazards such as dust, noise, and flies would also be carried away.[43]

Nonetheless, as the architect Shmuel Mestechkin explained, above all, the gardens in kibbutzim reflected Jewish European orientalist fears of the desert, dust, and drought, and their longing for familiar European landscapes.[44] Indeed, the gardens, especially those located in the Judean and the Negev deserts, were usually distinctively different from the landscapes in which they were embedded. Moreover, the architect Freddy Kahana notes that, during the 1940s, the planning programs of most kibbutzim were identical, and did not consider the local climate or other local environmental elements.[45] As Muky Tsur and Yuval Danieli write, in Palestine, architects and planners "did not protect the green lungs, they were forced to create them."[46]

In conclusion, the garden city was a popular method of urban planning for residential communities of the early twentieth century, which aimed to combine urban amenities and green zones while also providing improved hygiene standards. Jewish planners in Palestine adopted this scheme and implemented it in the first plans for Tel Aviv and in other towns and settlements as early as the 1910s and 1920s. However, while the concept of the garden city was developed in Europe as a response to the perceived urban ills of the Industrial Revolution, the garden city in Palestine was transplanted to a preindustrial setting, and served largely as a means to cope with the local climate and environment, as well as to demonstrate the Western cultural identity of the city's new inhabitants.[47]

HOW TO SAY "CLIMATE" IN ARABIC? JEWISH RESIDENTIAL SPACES, 1910S-1920S

In 1907, the Hilfsverein der Deutschen Juden (Association of German Jews) decided to establish the Hebrew Reali high school and the Technikum Institution for Engineering and Technology in Haifa. For this significant and costly project, the association wished to hire a Jewish architect, and in 1909 the job was offered to Alexander Berwald, a German-Jewish architect who worked at the time in Berlin for the Prussian government. Berwald accepted the offer, but before commencing the planning work decided to visit Palestine to study its local conditions. During his journey, Berwald visited Jaffa, Acre, Jerusalem, and Haifa, where he documented various local Arab houses in pictures and studied their building methods. Following his return to Berlin in 1910, he published an article in the ZO's newspaper, *Die Welt*, entitled "Bauliche

Probleme in Palästina" (Building Problems in Palestine), describing in detail the principles of local Palestinian architecture.[48]

Few Jewish architects at the beginning of the century gave expression to the ideas and thoughts that guided them in the process of planning and building. By contrast, Berwald explained and contextualized his ideas before commencing his plans. In his article, he repeatedly emphasized the importance of adjusting houses to the local environment, noting that climate was the element with the greatest influence on local architecture. Berwald also expressed his appreciation for the traditional architecture and the ways in which it had adjusted itself to the local climate, "to the extent that in different climatic regions in Palestine, distinct building models are used, for instance, Jerusalem is different from Haifa."[49]

Berwald's article in *Die Welt* was one of the only texts written about local Arab architecture in the first decades of the twentieth century. Another noteworthy text on this subject was published more than twenty years later in 1933 by Tawfiq Canaan, under the title *The Palestinian Arab House: Its Architecture and Folklore*. Like Berwald, Canaan emphasized the role of climate in the development of local architectural forms: "Despite the abundance of cultures and empires that ruled this region in the past . . . the Arab house was hardly influenced by foreign architectural styles because the local climate had the greatest influence on its evolution."[50]

As already mentioned, Canaan was part of a group of Palestinian intellectuals who were busy documenting the folkloric traditions of the local rural culture in the first decades of the twentieth century. The historian Salim Tamari reminds us, however, that, alongside their wish to protect Palestinian culture from external influences, these intellectuals were also largely influenced by European Romanticism.[51] In other words, both Palestinian as well as Jewish accounts of the local built environment tended to present the Arab artisan as possessing a deep understanding of the natural environment in Palestine.

However, although some Jewish architects viewed Arab architecture as "natural" and "organic," others chose to appropriate its cultural roots, while arguing that its origins lay in biblical times. In an article published in 1927 in the professional journal *Binyan Ve-Charoshet* (Building and Manufacturing), the architect Josef Awin wrote:

We have come to settle a land of great tradition. In cities and towns, we find crafted houses that seem to have grown from nature, full of loyal tradition, which is . . . the continuation of our people's building traditions in ancient times. . . . This art of building not only fits nature but it also fits [local] building material and the climate of our country.[52]

Nonetheless, the so-called historical continuity of building methods in Palestine was not necessarily favorable to Jewish architects, who were most often unfamiliar with the country's natural conditions. Owing to this problem, one of Berwald's declared objectives was to help Jewish architects avoid making undesirable mistakes by introducing them to local traditional building methods. Berwald argued, for instance, that the reason for the construction of the thick walls in Arab houses (usually between thirty and forty-seven inches) was that they provided a local solution to the temperature differences between day and night inland, as thick walls were believed to provide thermal resistance. Still, the choice of building methods and materials was not always a consequence of their suitability to the local natural conditions. It was often also influenced by various technological, economic, and social limitations. For instance, an article in the Templer bulletin *Die Werte* from 1875 stated that thick walls in Arab houses were the result of the heavy weight of the traditional domed roofs, which required stable foundations.[53]

In response to the *ḥamsin* and its effects on houses' interior spaces, Berwald argued that rooms should be organized so that the air could always flow between them. He claimed that "this principle is implemented successfully in Palestine."[54] For the same reason, Berwald also recommended planning houses with high ceilings, as this was believed to relieve the heat in rooms and, moreover, to install hatches above the windows to enable the warm air to dissipate during hot summer days.[55] During the late nineteenth century, glass was relatively rare and expensive in Palestine, and windows were often closed with shutters made of wood or metal. Berwald stressed the need to enable maximum ventilation into houses and advised against closing the shutters. However, he also acknowledged that leaving the shutters open would heat the room as a result of direct sunlight. As he explained, the local solution to this problem was manifested by the *mashrabiya,* a window enclosed with carved wood latticework that prevented strong sunlight but enabled light and air to enter

Figure 8: Two traditional ways of handling the Palestinian heat: hatch above window (a) and wooden lattice *mashrabiya* (b). Courtesy of David Kroyanker.

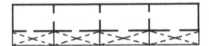

Figure 9: Row of rooms and arcade corridor as described by Alexander Berwald, 1912–13. Illustration by author.

rooms. The Arabic word *mashrabiya* means, among other things, "the place where the water is cold," denoting a place protected from sun but well-ventilated. The lattice-style window was an important component in the Arab house, moreover, because it enabled sufficient separation between private and public space.

Although the purpose of Berwald's journey was to study the local conditions for the planning of a public building, most of his descriptions referred to residential houses. In this context, he also discerned a few housing models that in his eyes perfectly suited the local climate. His preferred option was the one-floor plan with a row of attached rooms. Each room had a door that led to a long and covered corridor. Berwald claimed that this model enabled a very good flow of air and implemented it on the ground floor of the Technikum building.[56]

During the 1920s, around forty thousand Eastern and Central European Jews arrived in Palestine. Among them were many architects, including Josef Berski, Judah Megidovitch, Josef Berlin, and Josef Minor. Because the written materials produced by this group were often not

Figure 10: Alexander Berwald,
drawing for the Technikum
building, 1912–1913.
Courtesy of the Historical
Archives of the Technion in
memory of Nesyahu, the
Central Library in memory of
Elishar, The Technion—Israel
Institute of Technology.

preserved, it is difficult to assess whether they discussed and incorporated climatic and other environmental considerations into their architectural plans. What can be said with confidence is that, like Berwald, these architects regularly utilized oriental motifs in the buildings they designed. However, the adoption of oriental characteristics by the architects of the 1920s was not entirely the result of their encounters with Palestine's architectural morphology. In fact, this trend had already begun during the late nineteenth century in Europe and North America, when synagogues and other Jewish institutional buildings were designed with oriental attributes such as domes, minarets, horseshoe arches, and ornaments with plant decorations, geometrical forms, Hebrew letters, and psalms from the Bible. Unlike Berwald, while the architects of the 1920s embraced various nuances from the Muslim and oriental architectural order, they often dismissed local Arab architecture as primitive. For instance, during the construction of the Herzliya Hebrew Gymnasium, a building with clear oriental motifs, when the contractor complained about the complexity of the architectural plan, he was told that he should have known that the Gymnasium "would not be anything like a simple Arab house."[57]

The oriental style of the 1920s not only reflected the multifaceted Jewish identity in Europe and in Palestine but also mirrored European colonial architecture. The meeting of the British and the French empires with Asian and North African architecture and morphologies often led to the adoption of local aesthetic motifs, which were then intertwined with Western architectural building methods and techniques. This new hybrid style, often termed Eclecticism in the professional literature, was popular

Figure 11: Herzliya Hebrew Gymnasium. Courtesy of the Central Zionist Archives.

in Europe at the turn of the century. According to the historian Çelik Zeynep, however, studies on colonial architecture tend to emphasize the export of architectural models from Europe to the colonies, while neglecting the contribution of colonial control to the forming of architectural models within Europe itself.[58]

Most of the Eclectic style's characteristics traveled back and forth between metropolises and colonies, subsequently creating this hybrid style. However, in the case of Jewish architecture this amalgamation of knowledge and stylistic features was eventually transported by Jewish European settlers to Palestine, thereby giving it another colonial and transnational dimension. Specifically, the adoption of Eclecticism by Jewish architects assisted them in creating a unique settler architectural language that reflected both their occidental and oriental background.

Like Berwald's preferred plan of the row of rooms, flats in Eclectic houses usually included several equal-sized rooms organized in a row, with one room leading into the other. In addition, Eclectic flats often had a corridor that provided each room with another entrance or exit. This nonfunctional organization of the internal space—which may also have originated in nineteenth-century European middle-class flats—gave tenants the freedom to decide which room should be used for which purpose, thus also allowing them to make their own decisions in accordance with the wind and sun directions. The fact that each room had at least one window and two doors (one to the corridor and one to the next room) allowed beneficial ventilation within the flat. Furthermore, the ceilings of flats in the Eclectic style were usually about thirteen feet high, which was believed to cool the indoor space. In addition, most of the windows in Eclectic houses were large and vertical and often had hatches.

During the 1930s, the main criticisms of the Eclectic style in Tel Aviv derived from modernist architects, who blamed their predecessors for focusing exclusively on the houses' façades and not giving sufficient thought to the organization of the rooms and their relation to air and light. As we will see in the next section, room and building orientations were key modernist issues in dealing with the local climate. Modernist architects viewed the uniform functions of the different rooms in Eclectic houses as the ultimate proof of the style's incompatibility with Palestine and the inability of Eclectic architects to understand the local environment. However, the inhabitants of Eclectic houses did not always agree with this criticism. One contemporary wrote:

> Back in the day [1921–1927] when a man wanted to build a house and turned to an architect for a plan, the latter remembered that he was expected to plan the house in Tel Aviv, which is located in the lowlands at sea level and its climate during most of the [year] is very difficult—owing to the great heat and extensive sweat, he would plan the house in respect of the local climate.[59]

The writer further described the ways in which the architectural features of Eclectic houses—especially the window hatches—alleviated heat in the rooms and provided tenants with a convenient microclimate.

Both Berwald in the 1910s and the Eclectic architects in the 1920s aimed to express a collective identity in their architectural work, reflecting their perceived autochthonous belonging to the country. To them, this combination meant complying with local architecture to different degrees, as well as studying its methods of addressing climatic and environmental issues and at times even appropriating them. Although they also imported Western architectural ideas and methods, their buildings eventually reflected a complex exchange in which indigenous knowledge was central to shaping the settler culture. Jiat-Hwee Chang and Anthony D. King describe a similar process of combining local and metropolitan architectural models in British colonies, which they term "transculturation."[60] Nevertheless, the absorption of local knowledge by Jewish architects was short-lived in Palestine. With the advance of the International Style, attempts to study the local building methods were gradually relegated to the side and eventually replaced by Western scientific solutions to alleviate the heat in Jewish houses.

LOCAL HEAT — INTERNATIONAL STYLE

The prevailing environmental conditions in Palestine often led to confusion and uncertainty among Jewish architects in Palestine and required the adoption of new approaches to construction in order to relieve the heat in the houses for the convenience of the Jewish European settlers. However, instead of taking advantage of indigenous experience and knowledge—as some Jewish architects had attempted in previous decades—the architects of the 1930s and 1940s preferred to study the local climate and create their own solutions by implementing Western scientific building methods.[61]

This tendency was not unique to the Yishuv, and it can be traced in other European colonial enterprises, especially in the tropics. Chang and King discuss the development of tropical architecture,[62] which, they argue, reflected shifting colonial discourses on medicine, climate, and race. They identify at least two key moments in this development. Before the nineteenth century, British colonizers in India tended to absorb and implement indigenous forms and concepts of architectural adaptations to climate. However, in early-nineteenth-century India, the former "reverence" for indigenous knowledge eroded, and British architects moved toward metropolitan architectural models designed to protect Europeans from warm climates and their supposed miasmic threats. These architectural models were also designed to create and maintain the visible separation between colonists and indigenous people.[63]

Chang and King argue that the emergence of strong professional associations during the early nineteenth century was one of the main reasons for British architects' neglect of earlier attempts to imbibe local knowledge in building techniques and develop new systematic and technoscientific ways to confront climate.[64] Likewise, during the 1930s and 1940s in Palestine a central group of modernist architects called "the Architects' Circle"(Chug Ha-Adrikhalim) was altering the architectural discourse. This group was officially established in 1932 by several up-and-coming architects such as Arie Sharon, Josef Neufeld, and Zeev Rechter, who stated their case eloquently in journals and magazines of their own publication. In addition, the group members' close connections with the leadership of the Yishuv and their commitment to modern-progressive values facilitated the official adoption of their style in Palestine. Within a few years they occupied key municipal positions and were involved in shaping building regulations across the entire country. These architects were considered to be at the cutting edge of the profession and played a key role in the prosperity of Palestine, which at the time was driven mainly by the construction sector. Indeed, their impact and access to resources and power were so significant that even those architects who had not been formally trained in the modernist schools in Europe soon began planning identical buildings.[65]

From the very outset the migration of modern architecture (also known as the International Style) to Palestine was predicated on its foundational ideas of functionality, internationality, and progress, which were closely associated with the Zionist aspirations of creating a "new" and "healthy" Jewish society in the country. The principles of the International Style, which made it possible to copy and transpose residential models from Europe to other countries, were based on beliefs in a "rational" and "productive" society that was supposedly free from local, historical, cultural, or religious influences.

According to Mark Crinson, modern architecture's break with national conventions and European traditions and hierarchies has given it the appearance of an anti-colonial style within the historiography. However, Crinson claims that modernist architects hardly ever voiced opinions against colonialism. On the contrary, he writes, "perhaps it might be better to speculate that where modernism was not a disavowal of imperialism, it was actively deployed as a way of improving the functions of the colonial city, treating the colonies as a laboratory of modernity, elaborating the rhetoric of hygiene and health and the dualistic narratives of traditional and modern, even epitomizing the benevolence of the West."[66] The architecture critic Sharon Rotbard furthermore emphasizes the colonial attributes of the International Style by drawing similarities between the architecture of Tel Aviv and Algiers: "Both were attempts to found new, modern European cities overlooking or besides what were perceived as ancient crumbling dwellings. Both were exercises in settler colonialism and both were called White Cities."[67]

From 1934, the Circle published a journal initially titled *Building in the Near East*, and later simply *Building*. Climate was discussed at length in *Building* and similar architectural platforms, and the many articles published on the subject were written by renowned architects and phrased in the forms of guidelines and recommendations for the beginner architect in Palestine. In this context, architects often described climate as "a cause for concern for the architect,"[68] an "obstacle,"[69] and a "problem."[70]

One of the most important architectural solutions for the problems presented by climate was the positioning of living spaces within the flats in accordance with prevailing winds. In fact, the discussion of wind directions was so central to the modernist architectural discourse that it led to different orientations of flats in places with different topographical and climatic characteristics. For instance, while in Tel Aviv the prevailing winds blew from the west during the day and from the east at night, in Haifa the Carmel Mountain range blocked the eastern wind from reaching neighborhoods to the west of the range. This led to the design of different floor plans in each city.

The standard model for the Haifa flat determined that the living rooms, bedrooms, kitchen and balcony all faced west to enable maximal ventilation. Other rooms in which tenants spent less time, such as the stairwell and bathroom, faced east.[71] In Tel Aviv floor plans, on the other hand, the bedrooms faced east while the kitchen, living room and balcony— spaces used by tenants during the day—faced west, in order to receive as much wind and as little sunlight as possible until the afternoon. The architects preferred this model because it created a correlation between climate and the desirable and conventional working and sleeping hours. Karmi praised the Tel Aviv model, writing: "In the morning, the bedroom [facing eastward] does not have enough time to become heated up and cools later during the day, and that's the important thing. This allows for sleep and rest after the working day. . . . The east is essential for sleeping."[72]

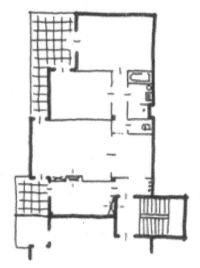

Figure 12: Shlomo Ginzburg, sketch of typical Haifa apartment floor plan, 1936. Clockwise: bedroom, bathroom, entrance, kitchen, living room, another bedroom. Courtesy of the Ginzburg family and *Building in the Near East*.

In general, the functional arrangement of rooms in modernist houses was supposed to promote the effectiveness and productivity of the tenant. Rooms whose typical activities were related were usually located next to each other. Thus, for instance, the kitchen was usually located next to the dining area.[73] More importantly, however—and as Karmi's words indicate—the decision to have the bedroom face east was justified not only in terms of the wind and sun directions desirable to minimize heat, but also based on the assumption that in the morning the tenant had to wake up early and start the working day. On the other hand, by the afternoon when the tenant was supposed to return from work, the room would have cooled enough to allow rest. Orienting the rooms based on the sunlight angles at each hour of the day framed their spatiotemporal location and, consequently, the activities that tenants were supposed to perform during the day and night, respectively.

Particular climatic importance was associated with the hallway, whose recommended orientation in the Tel Aviv flats was between east and west, in order to enable ventilation both during the day and night. Karmi identified the hall as a particularly useful space and praised its multiple advantages. He wrote:

> The lighted hall is used in the apartment as a dining area and playroom, the baby's area if necessary, or the working and sewing area. But its main func-

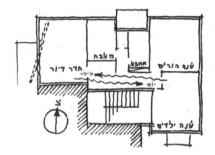

Figure 13: Dov Karmi, sketch of typical Tel Aviv apartment floor plan, 1936. Counterclockwise from below: children's bedroom, parents' bedroom, bathroom, kitchen, living room; serpentine arrows indicate day winds (below) and night winds (above). Courtesy of Ada Karmi and *Building in the Near East*.

tion is *to enable cross ventilation* in the flat. It is naturally desirable to place the lighted hall near the kitchen. . . . Wherever the lighted hall is located next to the kitchen and allows ventilation . . . flats are optimally organized and ventilated and family life is centered around the hall, which opens the flat to the movement of the winds.[74]

Locating the hall next to the kitchen reflected the importance attributed to hygiene in the architectural and medical discourse of the time, articulating the emphasis it placed on the interrelatedness of cleanliness, ventilation, and health. Moreover, discourse on climate and health in residential spaces was not limited to architects alone. Physicians and meteorologists were also involved in determining the conditions for the desirable Jewish flat in Palestine, and in so doing they often cooperated with architects. Cooperation, however, did not prevent criticism of contemporary architectural practice by the former.

For instance, the importance attached by modernist architects to the ventilation of flats was often denounced. The physician Walter Strauss argued, for example, that "a system that makes the wind the main element of the flat cooling is fundamentally wrong." According to Strauss, the key reason for what he described as the "harsh [thermal] conditions" in newly built flats was a "result of building malfunctions" and an indifference to the effects of solar radiation, rather than being "an inevitable outcome of the country's climate."[75]

From the 1940s, a greater number of scientific experiments were conducted on the effects of climate on buildings, many of which focused on solar radiation and its impact on different building materials. In 1942, the German-Jewish architect Joseph Werner Wittkower conducted an extensive analysis of the effects of solar radiation on indoor spaces and the health of tenants, which he summarized in an unpublished manuscript entitled *Bauliche Gestaltung Klimatisch Gesunder Wohnräume in Palästina* (The Design of Climatically Healthy Living Spaces in Palestine). The introduction revealed Wittkower's strong commitment to scientific research, which he believed to be the only reliable and effective tool for achieving "healthy constructions" as opposed to what he termed as "historic methods of trial and error."[76]

However, despite the considerable effort devoted to studying local climatic conditions, modernist architects were continuously blamed for their desire to maintain a stylistic integrity that overrode the need to provide climatic solutions for buildings. Dr. Groschka, director of the Tel Aviv municipal hospital Hadassah, wrote that "the most important hygienic demand for building houses in our country is the summer heat, just as the most important demand in Northern European countries is the cold climate. Thus, whoever builds here just the same as in England, Sweden and Holland builds according to fashion."[77]

In fact, the replication of European residential models in Palestine was not a result of professional carelessness or indifference. As mentioned earlier, the foundational ideas of the modernist architectural style were functionality and internationality. Thus, although modernist architecture was grounded on technological and aesthetic standards that were biased towards northern cultures and climates, they were understood as essentially reflecting international functional needs.

In 1929, Frankfurt hosted the Congrès Internationaux d'Architecture Moderne (CIAM), the most important forum for European modernist architects. At this conference, the participants agreed on a series of criteria entitled the "Minimum Dwelling Standards," aimed at guaranteeing a family's physical and mental health in a flat and quantified in terms of the availability of fresh air and natural light.[78] The design of space based on a series of rules enabled a residential concept that could be transferred from Europe to other countries with only minor modifications. In other words, the minimum criteria were a kind of "mobile rule kit" or "instruction manual" (terms invoked in this context by anthropologist Ofra Tene), which professionals could take with them around the globe. Wherever they

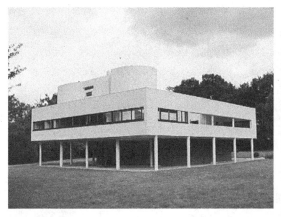

Figure 14: Curtain wall of the Bauhaus School, Dessau (a) and Le Corbusier, ribbon window in Villa Savoye, 1931 (b).

arrived, architects studied the objective environmental conditions and adjusted the CIAM rules according to their findings.[79]

The adjustments made by the Circle Architects to the International Style were often a direct application of the functional and international principles laid down in CIAM's 1929 Minimum Dwelling Standards. Nevertheless, climatic issues that were raised by direct encounters with the local conditions, and that were not predefined in the minimum criteria as potentially problematic, were often relegated to the sidelines of the professional discourse and did not lead to many practical adjustments even when the architects themselves believed in their importance.

When applying CIAM criteria, architects in Europe sought to maximize the amount of light in apartments. Accordingly, during the 1930s, they made frequent use of glass. Glass had aesthetic qualities but also a functional and economic aspect. Allowing the sun's rays into the building's interior through large glass windows and walls helped to light and warm the structure from within, enabling the minimal use of artificial lighting and heating during the daytime. The two main stylistic applications of this concept by European and North American modernist architects were the "horizontal ribbon window" and the "curtain wall," made entirely of large glass surfaces suspended in metal frames.[80]

Figure 15: Ribbon window adjusted to the local climate, 37 Lilenblum Street, Tel Aviv. Courtesy of Gabriele Oschatz-Cohen.

A modest version of the curtain wall technology found its way to Palestine in the mid-1930s. Owing to the hot summer, however, it was used sparingly for design purposes, such as emphasizing the vertical element of stairwells. Just as curtain walls were deemed unsuitable for Palestine, ribbon windows, too, were problematic in the sunny climate. However, architects refused to completely abandon this fashionable form. Instead of an elongated window stretched across the entire façade, they created a horizontal concrete slab, which was occasionally emphasized by a narrow awning in which small windows were installed. These windows were set at a maximal height of 6.8 feet above the ground to minimize penetration by the sunlight. In addition, narrow shade awnings were installed above the windows and balconies. Apart from their aesthetic resemblance to the European ribbon window, these awnings were supposed to provide shade in summer and shelter the windows against the winter rains. Nevertheless, their aesthetic significance often overrode their functional significance as awnings were, for instance, also installed on north-facing facades where there would be no direct sunlight.

Another way of quoting the ribbon window in Palestine was through the use of elongated balconies that highlighted the façade's horizontal section. The balcony's significance was primarily aesthetic, and only occasionally did the modernist architectural discourse of the 1930s and 40s refer to it in climatic terms.

In Palestine, the balcony had been a common architectural element long before the advent of the International Style. Already in the 1920s it had become an important and central space within the settler community's residential culture, owing both to the climatic comfort it offered and its unique character as a liminal space. The balcony offered some of the privacy and protection associated with home, while also presenting opportunities to interact with the street. During the twentieth century, its centrality to daily life was also widely articulated in urban poetics. But despite its popularity, the concrete balconies of Tel Aviv International Style houses were often criticized, as their shape prevented air from flowing into the house and, furthermore, they tended to quickly overheat. In 1935, a contemporary wrote:

> And the balconies! I set aside the question of architectural style. . . . In terms of the climate conditions in Tel Aviv, the balconies in the new houses—airtight concrete boxes instead of iron lattices—are really an outrage, that nothing but sheer idiocy could have blinded the architects to their disadvantages. Concrete boxes blocked against the wind, and on the other hand their large and continuous area may be overheated by the sun during the daytime. . . . And whoever sits on the balcony at sundown for a breath of wind feels as though he is bathed in steam, and must the architect be an expert in physics for this purpose? Where indeed has such elementary knowledge disappeared when balconies of this kind were allowed? It appears that this is not a financial issue. Or had stylistic preferences decided the issue? Stylistic integrity and public health—which is to be preferred? And our architects have determined that stylistic integrity comes first and people's health a distant second![81]

An additional hallmark of the International Style in Tel Aviv was the white color of the buildings, which eventually came to be associated with the city as a source of pride. This feature had climatic significance and was discussed extensively by contemporary physicians and climatologists. Nevertheless, it was almost entirely ignored in the architectural climatic discourse. The only architectural text I have located that refers to the choice of color was written by the German Jewish architect Julius Posener, and in any case was related to a completely different issue. According to Posener, the white house was devoid of historical traces and "free of past memories,"[82] rendering it suitable for the newly recreated Jew in Palestine. White was the trademark color of modernist villas built in this period in Europe. These were the buildings Posener mentioned when observing Tel Aviv's new houses ("The houses being built today are reminiscent of the modern German villas from the years 1925–1930").[83] Thus, for Posner, the question was not about climatic practicalities but about the symbolic dimension of the color white, which was culturally coded as "modernist."

Like the balconies, however, the public often manifestly disliked the choice of the white color. In 1933, for instance, the renowned painter and author Nahum Gutman wrote in the *Tel Aviv Municipality News:* "White radiates from within itself and the eye cannot bear it (this matter was commented upon several times by the ophthalmologist Dr. D. A. Friedman). The houses' colors require softening, darkening, and shading. But one thing is clear: the determination of colors must not be left to individual taste."[84] In addition, the anonymous *Ha-Aretz* reader quoted above did not mince words on this issue. Following their criticism regarding other features of modernist houses, they wrote: "The snow-white color— in our country . . . and when the bright sunlight shines on the white houses the place becomes a sea of light—and also of heat in the summer— which is really blinding. And we have no shortage of people with eyesight disabilities in our country as it is!"[85]

The Circle architects' faithful observation of the CIAM guidelines may explain the importance Jewish architects placed on studying wind direction, which came at the expense of the almost total neglect of other architectural factors that might affect the level of comfort within the residential building, such as the balcony and color of the houses. The next section of this chapter further shows the discrepancy between the emphasis placed on climate in the architectural discourse of the 1930s and 1940s and the practical solutions provided to address the issue. As Or Aleksandrowicz demonstrates, following the continuous climatic problems that modern architects encountered in the buildings which they designed, during the 1950s new methods were developed to counter-manage the problems created, for instance, by the extensive use of glass. These solutions, mostly ascribed to the work of Le Corbusier and the Brazilian architect Oscar Niemeyer, included the Brise Soleil shade and other creative shade solutions, and were likewise often condemned as aesthetic sophistries.[86]

CLIMATE, ROOFS, AND BUILDING MATERIALS

Throughout the first half of the twentieth century, the Arab house and its perceived aptness to the local climate continued to serve as a point of reference for Jewish architects even when they chose consciously to neglect it. Aleksandrowicz reminds us that while questions relating to building materials are often understood as "objective" engineering considerations aimed at ensuring stable and secure houses, throughout history different materials have become more desirable than others despite offering poorer quality and climatic reliance. Aleksandrowicz attributes such choices to economic, cultural, political and other deliberations, which eclipse concerns relating to the materials themselves and their suitability to the environment.[87]

Up to the beginning of the twentieth century, most houses in Palestine were built from available local stone (usually limestone, chalk, sun-dried silt, and sandstone). Local stone was often utilized for both the foundation of the house itself (that is, the house was built on an existing stone), as well as the construction of walls, stairs, and flooring.[88] The stone industry, which had been central to the entire building industry in Palestine for centuries, experienced unprecedented changes during the twentieth century. The growing number of Jewish settlements in Palestine and the accelerating building industry that developed as a result of it increased the production of stone in Palestine on the one hand, and created a demand for cheaper building materials on the other. The Jewish engineer T. Tinovitch, who wrote an article about the history of building materials in Palestine in 1945, pointed out that, while the Jewish building industry utilized local stone during its early years, it slowly neglected the stone for alternative building materials. The reasons for this negligence were primarily economic and political.[89]

Although new building materials began to arrive in Palestine in the late nineteenth century, the old stone strip mines continued to use old-fashioned, traditional techniques. The work of extracting the stone was performed manually, and the pace of production was accordingly slow. Under these conditions, there was no possibility of the local natural stone competing with the prices of the new industrial materials, owing to which it became an expensive commodity overnight.[90]

Another important reason for the gradual neglect of the local stone by Jewish settlers and builders was tied to the national motives of the Jewish settler society. During the first decades of the twentieth century the question of materials was deeply bound up with a central ideological question, namely, that of Hebrew Labor (*Avoda Ivrit*). From about 1904, Zionist ideology increasingly advocated the idea that the return of the Jewish people to its "homeland" also demanded a return to labor. Jewish landowners often could not supply work for all the Jewish settlers and, on the grounds of competition, sometimes preferred hiring Arab workers, who demanded a lower wage and were in any case better acquainted with stonework.[91] Thus, as part of its settler national ideology, the Zionist economy aimed to forcibly remove the local workforce, an act that, according to Gershon Shafir, allowed "the settlers to regain a sense of cultural or ethnic homogeneity identified with European concepts of nationality."[92]

Nimrod Ben Zeev similarly addresses the new building materials introduced into the local market during this period. According to him, "the Zionist project required not only Hebrew construction workers, but a Hebrew industrial apparatus and the capital to establish and support it."[93] Moreover, in practice, the greatest shift in the Jewish building industry's move away from stone to using alternative building materials occurred more from necessity than ideology when, during the 1936 Arab Strike, the Palestinian stone industry was temporarily unavailable to Jews.[94]

The alternative material chosen to replace the traditional stone was concrete. This material was not only easy to produce, owing to the availability of cement in Palestine, but it also enabled the Jewish community to become independent of Arab labor and industry and, therefore, it gradually gained great Zionist symbolic value. Building with concrete was understood as supporting the same cause as Zionist afforestation and harvest. The renowned poet Natan Alterman positioned concrete in a large spectrum of Zionist practices in his poem "Morning Song," published in 1934. The poem describes the local natural conditions—the burning sun, the dew, the desert, the swamps—and the ways to harness and civilize them—through harvesting, gardening, draining swamps and building with concrete ("We shall dress you in a gown of concrete and cement").[95]

In addition to economic and political settler-national considerations, it seems that architects and investors also preferred concrete and cement on account of their cultural value. Modernist architecture developed around the constructive possibilities of concrete and its ability to form a foundation for pillars and rafters that were independent of the walls as the main carriers of the weight of the construction. The qualities of concrete not only transformed the exterior possibilities of construction, but also enabled a different internal structure that could, for example, enable large horizontal windows.[96]

Nonetheless, the main disadvantage of building with concrete was its low durability in the local climate. According to the renowned architect Avraham Yaski, "in most of the houses that were built along the sea and which observed its humidity, the iron infrastructure swelled and after only a few years it blew up the concrete."[97] An article from 1967 stated, moreover, that "the problem is: how to prevent the quick wear of houses' façades as was the case here in the 1930s and 1940s, especially along the coast."[98]

In addition, the use of concrete increased the need for external insulation materials in order to reduce the heat, cold, and damp within the houses, as Dr. Groschka discussed: "We need good insulation. In the past this was achieved by very thick [stone] walls. New building methods . . . reduce the volume of material and require the search for new [insulation] solutions."[99] During the 1930s and 1940s, insulation materials such as tar, cork, sponge, stone, and light concrete were imported from Europe to compensate for the deficiency of new building materials that were not always suitable for the local climate. From the abundance of articles that were dedicated to this issue in professional journals, we learn that the urgent need for insulation materials created an unregulated market. An article published in 1935 in *Building in the Near East* stated, for instance:

> In the last few years, the market has been flooded with dozens of new materials that are being announced as the best solution for the insulation of heat,

wetness, and noise. In one of the last issues of this journal we have called for our readers' attention. . . . Not all products that are advertised as protecting from wetness indeed improved [insulation]; the same is true for heat and noise. . . . The research on building insulation can be complex and is not yet developed enough and some of our architects did not have a chance to become familiar with it abroad.[100]

Nevertheless, a study from 1944 demonstrated that, owing to the high price of insulating materials, most houses were eventually not insulated at all.[101]

Another way to alleviate the heat in houses was by considering the impact of solar radiation on the roof. Until the 1870s, Western travelers who arrived in Palestine encountered a scene of flat or domed stone roofs. The winter of 1874 marked the beginning of a shift in this respect. As a result of unusual heavy snow, hundreds of roofs collapsed, and even more were cracked. The Templers were the first to draw practical conclusions, and soon after this weather disaster, they began importing roof tiles to Palestine. They advocated the use of tiles as, according to them, tiles were not only better suited for managing precipitation, but also enabled the use of new building techniques and materials, since the light tile roof permitted the construction of thinner walls. In addition, they preferred its cultural-symbolic value, which was associated with the European rural landscape. The Jewish European architects who settled in Palestine during the first decades of the twentieth century also preferred the tiled roof because they were more familiar with its use and the building techniques it required.[102]

It was only after the modern style entered Palestine in the 1930s that the flat roof returned to fashion. In 1935, Posener described the irony embedded in the cultural meanings that were attributed to this architectural style. He wrote: "In Germany, they called this type of architecture—of flat roofs . . . 'an architecture of huts' that originates in our Oriental blood; and here, this architecture seems foreign and European."[103] Indeed, alongside the modernist admiration of steel, glass and concrete which had originated in industrial construction, various modernist architects also wrote about the influence of the stylistic language of Muslim architecture on their work. In Le Corbusier's book *Poesie sur Alger*, the renowned architect praised the oriental beauty of North African towns and cities. He wrote that "excellent architecture is square."[104] A year after his death in 1966, Le Corbusier's *Voyage d'Orient* was published. This book assembled different texts that he had written during his journeys, including his first encounter with Muslim architecture. In these texts and others, Le Corbusier often emphasized the perfection implicit in the square and modular Arab geometry, which in his view amplified monumental buildings, thereby creating what he described as a "holy" proportion.[105]

In addition, the Zionist Institute for Building and Techniques in Palestine (*Ha-Mosad Le-Cheker Ha-Bniya Ve-Ha-Tekhnika*), in cooperation with the Meteorological Department at the Hebrew University of Jerusalem (headed by Ashbel), conducted research in 1944 that asserted that the domes in Arab houses contributed to reducing the heat in the interior spaces of the houses since the sun's rays heated different parts of the roof at different times of the day.[106] As we have seen, modernist architects in Palestine often distanced themselves from the local building traditions and as a result, they completely ignored the domed roofs of Palestinian houses, despite their own research on their climatic efficacy.

At the same time, some architects advocated the tiled roof for similar climatic reasons. In his 1946 article on the desired roofs for industrial buildings in Palestine, Wittkower wrote that "the best means . . . for the efficient cooling of the building during the night hours is any non-insulated roof that is made of one single layer. Every roof made of tiles or metal sheet is good for this purpose."[107] Despite contrary opinions concerning the suitability of the tiled roof for the local climate, modernist architects continued to design flat roofs in the houses they planned. Like other architectural elements discussed above, the choice of the roof was more likely related to aesthetic concerns than to functional ones. This feature was not unique to the International Style in Palestine. To illustrate the priority given by modernist architects to stylistic features over considerations relating to physical environmental conditions, the architectural historian and theoretician Reyner Banham examined the effectiveness of the flat roof in modernist buildings in Western countries. Banham found that most flat roofs in Europe suffered from leakage problems in the 1920s and 1930s. This fact was well-known to all the architects who built in this style, but it did not prevent them from continuing to plan flat roofs. This evidence led Banham to conclude that European architects did not rely enough on personal experience in the field, but largely on the dictates of fashion.[108]

As we have seen in this chapter, the issue of climate was central to the Jewish architectural discourse in Palestine throughout the first half of the twentieth century. Until the late 1920s, Jewish architects were generally more interested in exploring Palestinian ways of coping with the local climate, and as a result they often focused on the suitability of artisan building methods to the climate in Palestine. Nevertheless, despite their desire to appropriate local building motifs while sometimes presenting them as ancient biblical ones, Jewish architects eventually tended to rely on modern Western techniques, even when they proved less suitable for the local climate.

Given the health hazards that were associated with the climate in Palestine, modernist architects made considerable efforts to study its characteristics. Nonetheless, in most cases, these studies did not contribute to the formulation of a new architectural style. Instead, modernist architects were keen to reproduce the formal idiom of the International Style with adjustments to elements pre-designated as potentially problematic in CIAM's Minimum Dwelling Standards. Additional architectural and engineering elements that affected the microclimate in residential spaces, such as the shape of the roof and construction materials, were usually sidelined in the professional discourse despite the architects' awareness of the problems this involved.

Thus, despite the importance of houses in determining the daily thermal comfort of their tenants, the position of architecture between art and science enabled Jewish architects to interpret the scientific evidence concerning climate in selective and flexible ways, and simultaneously encouraged nonprofessionals and professionals from other fields of expertise to intervene in the architectural discourse.

4. Climate and the Study of Plants

Since the eighteenth century, natural philosophers in Europe had been debating what they called the desiccation theory, and North Africa and the Mediterranean often served in this context as popular sites for European pondering the potential of climate change. The more reliable the meteorological data became, and the more frequently statistics were used to analyze it, the greater were the speculations concerning desiccation theory. Besides meteorological and geological findings, arguments supporting desiccation theory were often also based on "climate reconstruction," a historicizing of climate that depended on ancient literary accounts. Ancient Greek and Roman texts were occasionally interpolated alongside archaeological findings and geological analyses to prove that climate change had occurred in Greece, Cyrenaica, and parts of Europe.[1]

Nevertheless, based on the number of scholarly accounts dedicated to it, the Hebrew Bible appears to have served as the most important textual evidence for climate change in both Palestine and Egypt. The American geographer Ellsworth Huntington, a vehement environmental determinist and one of the most prominent advocates of desiccation theory at the beginning of the century, wrote in 1911 that "in no other country could [climate] theories be so well tested, for [Palestine's] known history extends back to remote antiquity."[2] Later in the same article, Huntington added that during his expedition to Palestine he was "surprised also to discover how closely historical progress or decline appear to have synchronized with the changes in climate."[3]

Similarly, in an article titled "Is the Earth Drying Up?" published in 1914, the British geologist and explorer John Walter Gregory aimed to confirm climate change by comparing Palestine's contemporary physical conditions with its glorious biblical representations:

> Palestine is described in the Bible as a land flowing with milk and honey; it was fertile in vine and olive; many incidents suggest that it was well wooded; and, according to the Old Testament statistics, it had a dense population. Now it is a barren, arid land, with a scanty vegetation, swept by parching winds from the eastern deserts, and occupied by some 700,000 people, who are mostly Arabs and mostly paupers.[4]

Such accounts, obviously, obliterated the biblical stories that depicted droughts and famine, such as the story of Jacob, who left Canaan with his family and fled to Egypt once he realized a drought was coming, and Joseph, who advised Pharaoh to stock up grain as he accurately predicted the coming of seven lean years.

The scholarly discourse on desiccation theory in Palestine mainly occupied British, American, French, and German colonial experts. Among the international scholars who debated the question of climate change in Palestine, many were, in fact, Bible specialists. The arguments put forth by these scholars were usually based on archaeological findings and literal biblical descriptions of the Land of Israel, especially those that referred to the material dimension of ancient woods and forests, water levels in lakes and seas, and demographic diminution.[5]

Most of these scholars agreed on the reality of changes in the local climate but remained undecided on a few fundamental issues. Some scholars believed that desiccation was part of a long-term geological progression, while others thought it was a process that could be dated to more recent historical times. In addition, this discourse was concerned with the causes of desiccation, which were perceived to be either natural (autogenic) or a result of human interference (anthropogenic).[6]

Those who argued for anthropogenic explanations claimed that Palestine had never lost its fruitful potential. For them, the only reason for the possible "worsening" of its conditions was local Arab and Ottoman neglect ("*Unkultur*," as one German scholar described it).[7] These scholars alluded to biblical weather patterns in the Land of Israel, while stressing their continuation in Palestine. For instance, the German scholar Heinrich Hilderscheid, who conducted a comprehensive study on annual rainfall trends and seasonality indices in the Bible, pointed out that as in modern times, summer rains during the biblical period were extremely uncommon. Another supporting example provided by Hilderscheid was related to the wind direction in the country. He argued that in biblical times rain clouds likewise came from the sea coast and moved inland (west winds), whereas winds from the south and the east indicated dry weather (*hamsins*).[8] The head of the GPA, Max Blanckenhorn, similarly wrote that "nature has not changed [in the Orient] but rather the people have uselessly lost their temper there."[9]

Zionist scientists occasionally also participated in this international discourse, thus manifesting anew their part in endorsing the link between Judeo-Christian beliefs and colonial aspirations. When doing this they similarly adhered to anthropogenic explanations, clearly intended to present the "return" of Jews to Palestine as a project of redemption that would, among other things, rescue the land and its climate and environment from local mismanagement. Aaron Aaronsohn, for instance, wrote: "Thanks to the extraordinary fertility of their soil, these regions were the cradle and center of the great civilization of antiquity, and scientists agree that they have lost none of those qualities which then constituted their fruitfulness. Their present economic inferiority is entirely to be ascribed to the political and administrative system to which all these countries are subjects."[10]

Blaming the Ottoman Empire and Palestine's indigenous people for mismanaging the natural resources in their country was a common colonial practice. As mentioned earlier, Muslims in North Africa and the Middle East in particular were frequently blamed by European colonialists as the latter evoked what Davis terms the "environmental imaginary":

a form of orientalism that viewed the Middle East and North Africa as strange and "defective" environments that were on the verge of disaster. Generally, the local Arabs were blamed for this situation, accused of instigating deforestation and overgrazing, while Westerners were able to portray themselves as those who had come to "protect," "improve," "restore," and "normalize" these regions.[11] In other words, during the late nineteenth century, deserts and semi-arid regions were usually not understood as natural ecosystems in their own right, but rather as deficient ecologies that had come to exist in their current state owing to general neglect. According to Caroline Ford, besides political and territorial objectives the discourse on local neglect in French Algeria also included a broad spectrum of anxieties raised by the settler society such as "hygienic issues, race and . . . the possible decline of both empire and 'civilization' itself."[12] Thus, desiccation processes were not only seen as a peril to ecosystems and to "life on the planet" but also as a danger to the progress of European civilization.

One of the most frequently utilized colonial methods for solving desiccation, "improving" local climate, and introducing civilization into arid regions was afforestation.[13] The link between forests and climate change, which prevails to this day, dates to the writings of the Greek naturalist Theophrastus and remained relevant throughout most of the following millennia. Richard Grove claimed that the modern evolution of this link came to occupy Westerners through their experience in tropical regions in the early modern period. Nevertheless, a scientific approach to forests only developed in the mid-eighteenth century in Germany. During this period, forests lost their connotations of wilderness and became what historian James Scott has described as bureaucratically measured and calculated entities intended, first and foremost, for production and profit.[14] The French were the first to implement these new approaches to forests in their overseas territories, soon to be followed by the British, who practiced them imperially on unprecedented scale. At the peak of this process during the late nineteenth century, one quarter of the British Empire's lands were declared protected forest. Moreover, fifty separate British forest services were established in the colonies in only a few decades.[15]

As we have seen in the fields of expertise discussed earlier, colonial forestry ideas and practices naturally travelled with Zionist European settlers to Palestine. However, French colonial forestry was quite literally implemented in Palestine, as the first Jewish force to afforest it on a large scale was the French-Algerian administration of Baron Edmund de Rothschild, who supported the activities of the first Jewish settlements in the country starting from 1882. Aaronsohn, who was part of this administration, wrote that "the Hebrew colonies [in Palestine] are just the same as the French colonies. The colony's managers, besides very few of them, are French or obtained their education in France."[16]

Indeed, the Jewish forestry experts who worked in Rothschild's settlements often acquired their advanced studies in France. Besides Aaronsohn, whose biography we have already encountered in the introduction to this book, forestry experts who studied in France included prominent figures such as Josef Niego, Rachel Yanait (Ben Zvi), Zeev Meits, Shlomo Zemach, and Nathaniel Hochberg. During their studies, these experts usually focused on botanical methods developed in North African colonies that shared similar climatic conditions with Palestine.[17]

Following Niego's advice, the first large forest in Palestine was planted between 1895 and 1899 in Hedera. This project was financed by Rothschild's development agency and the Palestine Jewish Colonization Association (PICA) as a swamp-draining measure. The 250,000 eucalyptus seeds that were planted to create the Hedera forests were not brought from Australia, its native home, but rather from Algeria, where French settlers had planted large scale forests using this tree.[18]

At the turn of the twentieth century, it was believed that the natural aroma of the eucalyptus tree improved and refreshed the air. The tree's rapid growth was also believed to dry swamps and, therefore, to help fight malaria. In other words, because, as we have seen in previous chapters, the quality of air (often confused with the term "climate" on a local scale) was frequently associated with the emergence of disease, eucalyptus trees and forests were planted to cure the country of its so-called malign condition.[19] Sandler summarized the link between these ideas and explained their logic when he wrote in 1912 that:

The destruction of forests in Palestine had a baneful effect on the sanitary conditions of the country. . . . In a wooded district much of the rain is absorbed by the plants, which select some nutriment therefrom, and return water-vapor to the air. This increases the rainfall on the one hand, prevents excessive moistening or drying of the soil on the other. As a matter of fact, the plantation of rapidly growing trees on certain malarial districts of Palestine, on swampy soil, has either greatly diminished malaria or wiped it out altogether. Forests also purify the air.[20]

In addition to fighting malaria and desiccation, the act of planting trees in general, and eucalyptus trees in particular, was sometimes intended to provide more shade in Palestine. When meeting with Herzl in Jerusalem on a particularly hot day in November 1898, the German Kaiser Wilhelm II famously told the Zionist leader that the country was in urgent need of shade and water.[21] Following the planting of several forests and groves in the country, Alexander Aaronsohn (Aaron's brother), likewise wrote in 1916 that the "Eucalyptus [. . .] soon gave the shade of its cool, healthful, foliage where previously no trees have grown."[22]

While experts employed by the JCA and PICA were active in executing afforestation schemes of eucalyptus trees beginning in the late nineteenth century, tree-planting schemes within the executive institutions of the ZO were slow to materialize. Moreover, during the first years of this organization, its members were usually focused on the fiscal objective of tree planting. This objective was not so much centered on the timber industry, as was the case in other colonial settings, but—influenced by Warburg's professional experience in the tropics—rather on planting fruit trees.

Immediately after the establishment of the CEP in 1903, Warburg began developing a plan for planting profitable olive groves. Following his recommendations, the CEP developed the "Olive Trees Donation" association in 1904, which aimed to collect donations from Jews and Zionists across the world in order to fund the planting of olive trees in territories owned by the JNF. This campaign was meant, among other things, to provide Jewish settlers with labor and, it was stated, to provide the country with shade. The Zionist leader and chairman of the Board of Directors of the JNF in Germany, Max Bodenheimer, wrote that:

One of the most important departments of the National Fund [JNF] is the Olive Tree Fund, through which not only is the land rendered productive, but shady gardens are created where now sun-parched deserts extend, over wide tracts of the country.[23]

Figure 16: A JNF postcard depicting the certificate awarded for a donation to support the planting of five or more olive trees, 1913. Courtesy of the Leo Baeck Institute Archive at the Center for Jewish History, New York.

The establishment of the "Olive Trees Donation" association was, in fact, an important moment in the history of the JNF, as it laid the foundation for its tree planting activity, which was added to its initial activity of purchasing lands in Palestine. Yet it was to be several more years before the JNF crystallized and began to develop significant afforestation policies. This development included the transition from the initial planting of fruit trees into planting non-fruit-bearing trees. The reasons for this shift, made by Bodenheimer in 1912, were the extensive amount of time and effort it took to cultivate olive trees, and perhaps also the acknowledgement that competing with Arab farmers on the cultivation of this traditional local crop might turn out to be an economic disaster. However, the JNF's shift to non-fruit-bearing trees was also related to the ZO's increased interest in improving the local climate.[24] Bodenheimer wrote in his memoir: "When visiting the country, I became aware that the olive grove that has been idealized so often, was in fact a failure.... After consulting with Dr. Ruppin, I decided on a fundamental shift. Following my suggestion, eucalyptuses, cypresses, and pines were to be planted. These trees are meant to improve the local landscape and its climate."[25]

Another important element of Zionist afforestation was related to the biblical image of the land in Judeo-Christian tradition. As we have seen, European Jews and Christians in Palestine often understood the scriptures as sources containing an "original" description of the country, and inevitably these descriptions were compared with the settlers' encounters with the natural conditions and landscapes in Palestine. In this context, numerous contemporary scholars referred to the existence of lush woods and forests in Palestine in ancient times, which were associated with a better climate. The JNF even set itself the goal of renewing some of these specific forests. The author and Zionist leader Moshe Smilansky praised such activities. In a newspaper article published a few decades later he reflected on the contribution of the JNF in enhancing the desired biblical landscape in Palestine:

> The JNF has not only planted new forests but has also revived old forests, which were planted by our ancestors when they inhabited this country. These forests were abandoned, uprooted, gnawed, burned and lost during the evil and long years in which our people were spread over seven seas away from their homeland. Only a few wretched remnants were left [of these forests].... The area of ancient forests that was recently revived by the JNF covers 25,000 duman: [It includes] a pleasant forest in Samaria, northern to Nahalal; a few forests along Mount Ephraim, Mount Menashe and Mount Asher next to the new settlements; and the forest on the hills between the Zebulun and the Jezreel Valley.[26]

Yosef Weitz, the renowned director of the Land and Afforestation Department of the JNF from 1932 until the late 1960s, similarly wrote that the arrival of Jewish settlers in Palestine marked the end of an 1800-year human quarrel with nature. With the commencement of the Zionist enterprise in Palestine, "a new era begins in which man will stand beside nature, if not by protecting its existing state . . . then by planting and sowing for the creation of forests in the Holy Land."[27] Weitz also mentioned the biblical origin of the Hebrew word forest (Ya'ar), which primarily meant a stony ground that cannot be cultivated.[28] Through the explanation of this term, we are reminded not only of the centrality of the Hebrew Bible to Zionist popular and scientific thought, but also of the political roles that forests played as practical instruments with which to gain power over land.

From 1912, the JNF not only replaced the plantation of fruit trees with non-fruit-bearing trees, but also moved from purchasing only those lands that were suitable for agricultural settlements to purchasing stony territories, which were unsuitable for agriculture but could be used for planting non-fruit-bearing trees. This change in planting policy was intended, among other things, to enable the JNF to continue holding its lands under Ottoman rule, as the land reform of 1858 dictated that uncultivated territory would automatically be transformed into state property (*miri* land). JNF's policy remained relevant during the Mandate since the British law protected trees. In this way, the JNF became a leading force in the expansion of the "Jewish frontier" in Palestine by using trees as effective agents for a territorial cause.[29] Akiva Ettinger, one of the JNF's first agronomists and director of the Agricultural Settlement Department of the ZO, declared this intention explicitly when he wrote that the aim of forestry in Palestine was "planting many fast-growing trees that do not require long-term maintenance, and at the same time planting on lands that cannot be utilized for agriculture . . . such as rocky lands, swamps, and moving sand dunes."[30]

Weitz, Ettinger's assistant and successor, more clearly articulated this approach when he wrote that:

> If we would have relied solely on agriculture, in its economic meaning, the country would have been doomed to eternal neglect in two-thirds of its territory. However, luckily, man and agriculture have a loyal and devoted companion, the forest tree, which blooms the desert beyond agriculture by spreading its roots on stones and rocks, on sand dunes and swamps. . . . Thus, when the land cannot be used for agriculture, it can be used for forestation.[31]

During the 1920s and 1930s, the borders of this afforestation frontier mainly concerned the central and northern parts of Palestine, which were also the most fertile areas in the country. Within these borders the settlement campaign was gradual. Following the successful "redemption" of the Jezreel Valley, the marshy lands of Emek Chefer were next to be purchased and cultivated. This region was followed by a focus on several Jewish settlements in the Beit Shean Valley.[32] However, in the 1930s and 1940s Zionists also started to consider the Negev desert as a frontier in its own right while researching its climate, water, and soil conditions for agriculture and afforestation purposes.

Various Zionist expeditions had already explored this region during the 1910s and 1920s and the Jewish geographer Haim Bar-Daroma, for example, recommended that the Zionist movement invest more in this region. He wrote in 1935 that although the Negev had lost its commercial value since ancient times, it had, nevertheless, maintained its agricultural value, and due to its warm climate it could be revived slowly.[33] In 1939 Kibbutz Negba was established. This was the first kibbutz in the northern Negev, and it indicated the gradual Zionist interest in this geographical region. This settlement was followed by a few other settlements in the area such as Nir Am, Beerot Itzhak, Yad Mordechai, and Dorot. In 1943, following a big land purchase by Yehoshua Henkin and Hiram Danin, three further experimental settlements were founded in the northern Negev (Revivim, Beit Eshel, and Gvulot), which were supposed to study the conditions of the region and prove its potential fertility.

During this time, the Negev was presented as a geographical area that required cultivation and redemption, and one of the most popular suggestions for achieving this goal was afforestation. The relatively late attention given to the Negev did not allow for the accomplishment of large afforestation schemes during the Mandate. Yet these schemes became possible following the establishment of the State of Israel in 1948, when planting trees in the region increased year on year. For instance, the JNF planted the Lahav Forest in 1952 and the Yatir Forest, which covers 7,500 acres, in 1964. Both forests were meant to "combat desertification and heal the wounded earth."[34] Weitz wrote that the Zionist forests were not only able to fight off the desert but could also create a security zone for the people of Israel. Indeed, as he had already claimed in 1933, forests were instrumental in "determining geopolitical facts on the ground."[35]

As Novick demonstrates, the success of Zionist and British forests (before and after 1948) depended on the elimination of another local inhabitant—the black goat—which was seen as a grazing threat to young trees. British and Zionist officials believed that Palestine's "revival" depended "on a systematic shift in the order of the land."[36] However, these actions similarly posed a threat to the livelihood of Arab villagers and Bedouins, who often relied on the animal's products and milk as a main source of income. They saw the British restrictions as a tool to ultimately force Palestinian Arabs to sell their lands to Zionist organizations. In other words, in Palestinian eyes, afforestation projects and the collaboration between British governing powers and Jewish organizations became a tool for displacement.[37]

Indeed, during the Mandate period Zionist afforestation policies were intertwined with the local British forestry policy, which was concerned both with conserving existing shrubland and with planting new forests. Between 1921 and 1948, the British government planted 20 million trees on an area of 10,207 acres while supplying another 15 million trees to local municipalities. This mass afforestation, which was more than double than what the JNF achieved at the same time, was officially meant to improve the climate and stop the invasion of sand dunes as well as prevent erosion and collect water for underground aquifers.[38]

The so-called environmental decline referred to by the British was also addressed in the Afforestation Ordinance as early as 1920, that is, even before the official Mandate for Palestine was received. High Commissioner Herbert Samuel reported to the League of Nations in 1921 that "such are the first beginnings of a process which should add largely to the productiveness of Palestine, increase its rainfall and bring fresh charm to its scenery."[39]

When describing the cultural impetus behind Britain's afforestation schemes in Palestine, David Shorr writes that the British believed that:

> The cooler and moister climate produced by the forests would change the very nature of the countryside, rendering the landscape more pleasing to the European eye. The new conditions would be favourable to the raising of sheep and cattle, on the English model, instead of the destructive goats of the Mediterranean basin. Thus, would the Holy Land be changed—or returned—to one in which Europeans (whether Zionist settlers or British administrators) could live with no great discomfort.[40]

British afforestation also fulfilled an important political and territorial role. Forests were large land areas that were controlled and monitored by the government and in which grazing, woodcutting, and, at times, even loitering were prohibited in ways that crucially affected the local agrarian social structure. As Barton notes, "this was most likely to happen, not under a democracy where a majority of users—peasants, craftsmen, and traders—could block reform, but under an authoritarian regime where colonial overlords could, for better or worse, impose their control."[41]

ACCLIMATIZATION OF FOREIGN PLANTS:
SOURCES OF INSPIRATION

Prior to the British and Zionist afforestation schemes, tall trees in Palestine were very rare. During this period, the local flora mostly consisted of maquis and garrigue, shrubs or small oaks, and terebinth. However, as we have seen, because the colonizers were rarely familiar with the "original" flora of the country, they planted trees in Palestine that had never existed there before, thus radically intervening in local natural processes.[42]

Throughout human history, settlers have transferred seeds and plants from their homelands to their new destinations, thereby changing the ecology and landscape of their new homes in conjunction with their presence. In Australia, oak and elm were originally introduced by Europeans to enhance the timber industry. Similarly, in New Zealand and California, pines and cypresses (as well as some varieties of eucalyptuses from Australia) were planted to enhance the timber industry and to provide more shade. According to one source from 1956, for every 600 plant species in North America, only 100 were endemic while 400 had originally come from Europe and Asia.[43]

William Cronon writes that the settler-colonial landscape of New England was shaped by a variety of circumstances that did not necessarily correspond with the ecology of the region. Cronon asserts that settlers' descriptions of the natural environment usually contained comparisons with their former homelands, and that preconceptions regarding what the new country would offer often determined the kinds of natural information that was stressed and what kind was omitted.[44] Furthermore, John H. Elliott argues in a different context that "one of the paradoxes that run through the whole history of metropolitan-colonial relationships" was that colonialists, "even while coming to appreciate the qualities that made their environment unique, devoted a great deal of time and energy to making it resemble as closely as possible the environment that they had left behind."[45]

As the knowledge and scientific tradition of Jewish settlers in Palestine was tied to that of European countries, some of the plants and animals that were acclimatized in this territory came from Western countries. In the 1920s, the eminent position of the eucalyptus tree was replaced by the pine tree (*Pinus halepensis,* redubbed *Oren Yerushalaim* in Hebrew, meaning Jerusalem pine) which, as Braverman claims, has served ever since as an emblem of the Zionist project in Israel/Palestine. Zionist foresters liked the pine for several reasons. Pines grew relatively quickly and created a distinctly European landscape, thereby concealing the geographic dislocation of European Jews and reminding them of home. However, neither the eucalyptus nor the pine was particularly welcomed by the local Arabs, who named them *shajarat al-Yahud* (the Jews' trees). Once the trees grew, their needles formed a highly acidic ground cover that decomposed very slowly and made the forest ground sterile and inhospitable to further undergrowth and to most animal populations. Environmentalists coined the term "the pine deserts" to describe them. Today it is recognized that a monoculture of pines does not only prevent biodiversity but also increases fires, which spread more easily in non-diverse environments.[46]

Like the other aspects of nostalgia that were discussed earlier, for Jewish European settlers in Palestine the forest was a touching reminder of homeland landscapes, as is evident in many written accounts. For instance, the renowned poet Leah Goldberg wrote in a newspaper column from 1945:

> Young pine groves, who would not enjoy them? In this land they are a great wish for the future . . . but sometimes the heart is actually drawn to old trees. Those who have planted deep roots in the ground, their branches are convoluted, and their trunk tells of a permanent residency in this land. Trees that are not part of a project but are just present.[47]

As we have seen, when the local climate and landscape of Palestine was not compared with that of lost European homelands, it was usually compared with other similar climates where Western administrations had succeeded in managing profitable plants. This tendency was not unique to Zionists in Palestine. Brett Bennett writes that, while earlier forms of settlement and agricultural policies in the British Dominions had been motivated in part by ideas of climatic similarity between Britain and her new possessions, by the early twentieth century ideas of climatic equivalence were emerging between the colonies and dominions themselves.[48]

Similarly, Zionist scholarly attention often focused on modern agricultural and botanical knowledge developed in other settler societies. For instance, in 1928, Rachel Yanait Ben Zvi requested that the United States Forest Service in San Francisco send her information on planting trees in arid zones.[49] Warburg wrote that "the coastland of Palestine is at least as well suited for these cultures as the best fruit-growing districts of California."[50] Aaronsohn added, "Of all countries whose agricultural conditions may be compared with those of Palestine, only Tunis and California have agricultural schools where young farmers can successfully be trained in order to become useful experts in Palestine."[51]

Hebrew forestry and agricultural publications also frequently referred to scientific literature on the acclimatization of plants and seeds in other settler countries located in warm climatic regions. For example, the bibliography of *Cultivation of Sub-Tropical and Tropical Fruit Trees*, published in 1942 by the botanist Hanan Oppenheimer, mentioned the following titles:

H. K. Lewcock, *Pineapple Culture in Queensland*, Queensland Agr. J. 54.

J. A. Mitchell, *The Climate of Florida*, University of Fla. Bull. 200, 1928.

G. Ginestous, *Résumé de la Climatologie tunisienne*, Tunis, 1922.

H. V. Smith, *The Climate of Arizona*, Univ. of Ariz. Bull. 130, 1930.

Gr. Taylor, *Sea to Sahara, Settlement Zones in Eastern Algeria*, Geogr. Rev. 1939.[52]

Correspondingly, the content of foreign scientific journals, as well as the technological solutions they offered for overcoming climatic challenges, were continuously cited in Hebrew forestry and agricultural publications. One popular theme in these journals was the destructive impact of the *hamsin* on fruit plantations. These articles employed the knowledge of experts, who often compared the *hamsin* to the Santa Ana dry winds of Southern California or the Sirocco in North Africa, and who argued that European tree species were, in fact, less resilient to this type of weather.[53]

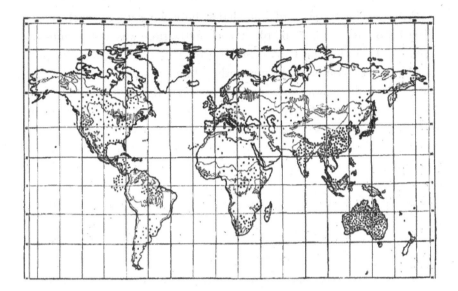

Figure 17: Israel Gindel, map showing the origins of plants in Israel, 1956. Courtesy of the Gindel family.

One of the solutions to this problem was the planting of trees to serve as windbreakers against the *hamsin*. This method had proved successful in the "Great Plains Shelterbelt," a windbreak project in the United States that began in 1934 and, that was intended to, among other things, prevent soil erosion. The project was initiated in response to the severe dust storms that occurred in the aptly named Dust Bowl, which had resulted in significant soil erosion and drought. The United States Forest Service believed that planting trees on the boundaries of farms would reduce wind velocity and lessen the evaporation of moisture from the soil. Weitz alluded to the American project when discussing a similar project in Palestine. He wrote that the windbreaking trees would protect humans, animals, and plants alike from the destructive impact of the *hamsin* winds and increase crop yield by 15 percent.[54] He argued that "with such scarcity of water, for the development of agriculture in the Negev, wind breakers are only second to water sources."[55] Windbreakers were also used to delineate borders of Jewish settlement, which according to Weitz were supposed to help protect them from their Arab neighbors.[56]

However, the transfer of knowledge between settler societies and Palestine on the subject of plant acclimatization was not a one-way exchange. In 1910, the United States Department of Agriculture published a bulletin composed by none other than Aaronsohn. The introduction to this bulletin was written by the acting chief of the United States Bureau of Plant Industry, G. R. Powell, who stated that the economic and agricultural importance of the Orient was beginning to be appreciated in the USA and that American explorers in Palestine had "learned in the Orient that there is every reason to hope that the vast arid and semi-arid regions of America can be rendered productive, although from the point of view of the European agronomist they would appear as worthless deserts."[57]

As mentioned earlier, following his pioneering exploration of emmer wheat, Aaronsohn was invited to tour the United States between the years 1909 and 1910. During his visit, he was warmly received by American experts, who also demonstrated their appreciation of his work in material terms. After returning to Palestine and establishing the agricultural experimental station in Atlit in 1910, Aaronsohn received farming machinery, an abundance of books and journals, and typing machines from his fellow colleagues and American donors. In return, the American agricultural ministry expected the Atlit station to provide it with dozens of local seeds and plants. In October 1911, Aaronsohn even went on an excursion to Upper Egypt initiated by the American Agricultural Ministry in search of selected dates to be acclimatized in the United States. Similarly, Aaronsohn corresponded with the Australian government about matters concerning afforestation and continuously provided information on tropical and subtropical plants to French colonial institutions in Tunisia.[58]

While Jewish acclimatization stations aimed to acclimatize foreign trees and plants such as the eucalyptus and pine, the knowledge they shared with agricultural experts and institutions in other settler societies was often based on their studies of Palestinian weather-resistant crops and traditional dry farming techniques. The agricultural station in Atlit, for instance, cultivated local figs, grapes, almonds, pomegranates, and other local crops, which were presented in the bulletin by Aaronsohn.[59]

Similarly, Mustafa Kabha and Nahum Karlinsky describe the protocols of the first cooperative of Jewish citrus growers, which demonstrate the extent to which these farmers relied on their Arab counterparts to learn the most fundamental knowledge of citrus growing and marketing. These protocols even reported how Jewish farmers obtained manure for fertilization from Arab *fellahin*, and also consumed Arab expertise on how to fight the diseases and insects that plagued the groves. According to Kabah and Karlinsky, this know-how considerably facilitated the increase in Jewish orange output per dunam.[60]

Thus, as we have seen in similar cases, when Arab knowledge was not rejected, it was usually appropriated. However, as we shall see in relation to some specific crops in the following sections, the traditional cultivation of local fruits was often presented via its biblical connotation, which charged appropriated Palestinian knowledge with a sense of Jewish nativeness. In this context many publications tended to present the biblical origin of local fruits, to which they added matching verses from scriptures.

However, while other settler-colonial enterprises used the biblical myth as a symbolic metaphor to create a sense of collective purpose for the ordeal of their experiences in uncharted territories, Zionist ideology presented the Hebrew Bible as a historical document describing the literal past of the Jewish people in Palestine. The first director of the acclimatization research station in Rehovot, the botanist Israel Gindel, preferred using the Hebrew term *imutz* [adoption] over the foreign term "acclimatization" [*aklimatisazia / iklum*], since, according to him, it was this term that was utilized in the Hebrew Bible to describe the same practice. In his book *Imutz Tzmahim* (Adoption of Plants), Gindel devoted an entire chapter to the ancient history of plant acclimatization in ancient Palestine. He began his account with cedar trees, which according to the biblical story were imported to Palestine from Lebanon to build the Jewish temple in Jerusalem; he went on to quote from further Jewish religious texts that referred to other acclimatized crops such as Persian dates, Lebanese garlic, Egyptian pumpkin, fava beans (*ful*), and lentils. Finally, Gindel quoted the Talmud, which described acts of acclimatizing plants in detail.[61]

The Bible was also used to draw similarities between Palestine and other settler-colonial projects. Despite the frequent exchange of knowledge with a variety of colonial enterprises, throughout the first half of the twentieth century comparisons between Palestine and California remained the most common. The American land conservationist Walter Clay Lowdermilk, for instance, who was sent to Palestine by the US Department of Agriculture in 1938 and was a member of the pro-Zionist American Palestine Committee,[62] wrote in 1944:

> Although Southern California is much larger in area and structure than Palestine, it is a remarkable counterpart in many respects of the Holy Land. The Colorado River with its muddy water is much like the Jordan. The Salton Sea, below sea-level and rich in salts, is much like the Dead Sea. The Mountains of San Gabriel topped with Mount Baldy remind one of Mount Hermon, and the San Bernardino Mountains remind us of the Jebel Druz of Trans-Jordan. The valley floor of the Santa Ana River is like the Maritime Plain of Palestine, in which irrigation is from underground waters.[63]

However, Lowdermilk did not confine his comparisons to environmental similarities between the two geographies. He argued that "colonization in America was like the colonization of Palestine—there were hardships and dangers in both cases."[64] Likewise, the agronomist and founder of the research station Rehovot, Itzhak Elazari-Volkani, described California as "an example and a role model" for Palestine,[65] and devoted an entire chapter of his book *Midot* to comparing Palestine and California. In the chapter, he discussed climatic and agricultural matters concerning both places, but also related their settler-colonial economic and political attributes and circumstances. For instance, he observed that both Jews in Palestine and Americans in California aimed to establish plantation economies that would rely on settler labor rather than on imported, cheap, and "colored" labor.[66]

One of the most important similarities between Palestine and California that enabled agricultural and economic prosperity in these arid regions was their massive reliance on advanced irrigation systems. Timothy Mitchell writes in a related context that technology "offered more than just a promise of agricultural development or technical progress. For many ... this ability to rearrange the natural and social environment became a means to demonstrate the strength of the modern state as a techno-economic power."[67] During the period examined in this book, the study of plant acclimatization became increasingly dependent on irrigation solutions to confront local natural conditions.

IRRIGATION TECHNOLOGIES

This section focuses on old and new irrigation technologies that were used by Arabs and Jews in the first half of the twentieth century in Palestine. Although water is usually considered an independent ecological entity which is not necessarily related to climate, it is also not entirely separate from it either. Within the field of agriculture, the availability of water was often associated with, and at times even dependent on, local precipitation levels. When rainwater was scarce, farmers needed to find alternative sources of water to nourish their crops. Moreover, the quality and amount of water required for this purpose was measured in correlation with other climatic conditions, such as the local attributes of different seasons, day and night temperatures, sun radiation, and humidity as well as with the quality and structure of the soil.

The availability of advanced irrigation systems was especially essential in British and Zionist afforestation and agricultural schemes, which often ignored local dry farming methods and instead chose to emphasize the acclimatization of various trees and crops that allegedly suited the local climate but, in fact, required very large quantities of water to prosper. Thus, this and the following sections present how Jewish experts dealt with water and precipitation scarcity and how they imagined, and indeed succeeded with the help of technology in engineering, "abundant" environments to suit their economic, political, and cultural aspirations.

According to Samer Alatout, while the British and other Western powers in the region generally tended to describe water as scarce, Zionists often aimed to describe the availability of water in a more optimistic manner. The only problem, according to them, was that water was "deficiently distributed." Water was often viewed as a resource that existed in abundance in biblical times (enabling the flourishing of the land of milk and honey) and that, like many other natural elements, only needed to be restored.[68] One of the biblical passages that n experts repeated in this context was Deuteronomy 7:7–9: "Behold, the Lord thy God giveth thee a good land, a land of water brooks and fountains that spring out of the valleys and depths, a land of wheat and barley, of vines, figs and pomegranates, of olive oil and honey, a land in which thou shalt eat bread without scarceness, thou shalt not lack anything in it."[69]

Zionist experts tended to emphasize water as the most important natural resource in the country and, thus, from the late nineteenth century onwards, they aimed to secure and increase the agricultural water supply using their material advantages over Arab farmers (such as foreign credit sources, educational opportunities, and access to advanced technology). For instance, during the Paris Peace Conference in 1919, in which the borders of the defeated nations of World War I were redefined by the Allies, Zionist leaders suggested future borders of Palestine that were based mainly on strategic water sources in the region. As Fredrik Meiton points out, in preparing for the Peace Conference, Zionist leaders, in fact, did not turn to the Bible or historians to delineate the desired borders of Palestine (as they would normally have done). Instead, they turned to the agricultural engineers Shmuel Tolkowsky and Aaron Aaronsohn to suggest a map of Palestine based on economic considerations.[70] Aaronsohn wrote in this context:

> In Palestine, like in any other country of arid and semi-arid character, animal and plant life and, therefore, the whole economic life directly depends on the available water supply. It is therefore of vital importance not only to secure all water resources already feeding the country, but also to ensure the possession of whatever can conserve and increase these water—and eventually power—sources.[71]

Water sources in Palestine were usually divided into two main categories—upper ground water and underground water. Traditional Palestinian agriculture used upper ground water in a rotational irrigation scheme called *fassil*. The *fassil* method was accompanied by a flooding technique that utilized gravity to irrigate lower spots, often resulting in the creation of pools or manmade swamps. In these terrace and terrace-like schemes, water distribution alternated from one area to another every twenty to thirty days. This method perfectly suited the agricultural practices based on seasonal crops, within which irrigation was only periodically required every year.[72]

At the beginning of the Mandate period, Arab agriculture was still generally dependent on rainfall. This dependency might also explain the type of crops cultivated by Palestinian Arabs, who focused mainly on dry cereals and on local fruits and vegetables. According to Roza El Eini, Amos Nadan, and others, from the late Ottoman Period until the 1940s grains were the predominant crops on Arab land, comprising about 75 to 85 percent of the total cultivated land, followed by olives and citrus. In addition, a two-field system was common on Palestinian agricultural land in which wheat and barley were grown as winter crops on one half of the field while the other half had a summer crop (e.g., sesame or millet) that depended on dew. In the following season, the second half grew the winter crop, while the first half was left fallow.[73]

Until the 1940s, Arab *fellahin* hardly invested in advanced irrigation systems. According to Nadan, this was mainly a result of the high price of such systems, which many farmers could not afford, as well as the cheap Arab labor that enabled irrigation by intensive manual work. Another issue that hindered Arab *fellahin* from investing in advanced irrigation was related to disputes and lack of clarity concerning the property rights on water sources. However, in the 1940s, when income from agriculture increased and labor costs in the Arab sector likewise rose, motor pumps became more common in local Arab agriculture.[74]

Nevertheless, in the years 1944 and 1945 55 percent of the Palestinian vegetable crops were still irrigated by dry farming techniques. During the same time, 95 percent of the Jewish vegetable crops were irrigated by artificial solutions.[75] Indeed, the general tendency of local farmers to adjust the type of crops they grew to the natural water regime at the beginning of the twentieth century was often understood by the Zionist and the British as backward and was, therefore, frequently criticized, as the latter wished to overcome the local climatic constrains by using advanced technology.

Upper ground water was also used by Zionist farmers who aimed to extract it from rivers. However, the number and level of exploitation of rivers in Palestine was relatively low.[76] Upper ground water solutions were not always easy to manage, owing to the geological and morphological characteristics of the country. The patterns of distinct rivers were also understudied during the first decades of the twentieth century, and therefore Jewish hydrology experts such as Shlomo Kaplanski, Joseph Breuer, and Judah Leo Picard were at first hesitant to initiate complex and expensive technological enterprises based on upper ground water.

The second type of water extraction during the first half of the twentieth century was underground water extraction, and it was usually accomplished through drilling wells. The mountains and hills in Palestine (especially along the Jordan River) serve as extensive aquifer systems. Rain that falls on these mountains not only evaporates or run off as surface water, but it also permeates the water table and then flows through the cracks of the underlying rock layers. Groundwater in the aquifer was considered to be a "safe yield" that could be pumped without adverse effects on the water left in storage.[77]

As has already been mentioned, before 1940 traditional drilling was limited in Arab plantations, and therefore water extracted this way could be used only in the immediate surroundings of wells and could not be distributed easily within large fields. It was only following the intensification of foreign agricultural activities (mostly in Templer and Jewish colonies) and, later, the use of electricity to operate water pumps, that water wells became an efficient means of irrigation. In the Templer colony of Sarona, a wind-powered pumping system was installed in 1879. In the 1880s, Edmond James de Rothschild helped the Jewish settlers of Rishon Le-Zion to perform deep drilling for water extraction, and a decade later two steam-powered pumps were installed at the wells of the wineries of Rishon Le-Zion and Zikhron Ya'akov. By 1897, internal combustion motors had been introduced into Palestine. Following these technological improvements, water in Jewish settlements could be pumped from a depth of 30 meters, compared with the traditional systems that only reached 12 meters. The water flow also increased from 5–8 to 20–40 cubic meters per hour, a change that enabled the expansion of the plantation areas. Between 1900 and 1930, more than a hundred wells were drilled along the coastline and the lower hills region, where the soil was sandy and the aquifer close to the surface of the ground. From the 1940s onwards, improved drilling technologies enabled the penetration of the stony ground along the mountain aquifer, which stored even more water.[78]

New research on underground water extraction has shown that the intensified use of this technology for agriculture eventually "results in rapidly falling groundwater tables, declining water quality, increased rates of saltwater intrusion and land subsidence, while also drying up natural water bodies like wetlands and rivers with detrimental effects to biodiversity. Excessive groundwater pumping may also irreversibly destroy or reduce underground storage capacity as well as damage hydraulic connections between surface and groundwater systems."[79] As we shall also see in the following pages, such research has also argued that colonial scientific ways of knowing (measuring, thinking about, imagining) and managing ground-water often created unsustainable models of farming and development.

Besides its ecological implications, irrigation in Palestine naturally also had a political dimension. Soon after the establishment of the British Mandate in Palestine, tensions began to mount between Jews and Arabs, resulting in violent encounters. Following the 1921 Jaffa riots, the British government published the Churchill White Paper, which had far-reaching effects on the Yishuv. Although the British still supported the principle of a Jewish national homeland in the country, in order to maintain the status quo, they decided to limit future Jewish immigration to what they defined as the "economic absorptive capacity" of the country.

This new provision, in addition to several years of below-normal precipitation, made the acquisition and development of water resources an even more important variable in the Zionist view. According to this view, the legitimacy of the Zionists' claim to a national home could be sustained only by extensive land holdings and a demographic presence approximating that of the Arabs. In addition, much of the land had to be turned into agriculture, and for this purpose large-scale irrigation was vital. From this perspective, the imperative of agricultural development was sustained not only by the Zionist ideology of transforming Jews into productive people, as is often emphasized in the historical literature, but also by a political "necessity."

Furthermore, the establishment of the Zionist water company Mekorot coincided with the Yishuv's separatist approach, which was enhanced in the aftermath of the 1936 Arab revolt. According to Alatout, while before 1936 water and water extraction were in the hands of private owners who managed it on a local and regional scale, Mekorot was established by the worker union *Ha-Histadrut*, the JA, and the JNF for the purpose of planning, executing and running waterworks for irrigation and consumption in the Jewish sector alone.[80] A national water company was also needed because the British made very little progress in advancing regional irrigation systems in Palestine until 1940. It was only in this year that the government finally established the Irrigation, Drainage, and Water Resources Service. However, even after establishing the service, no large irrigation schemes, such as those in Egypt or India, were employed in Palestine.[81]

One of the most challenging and exciting territories in Palestine, designated by Zionist and British for large agricultural schemes through irrigation starting from the 1940s, was the aforementioned Negev desert. The British Jewish businessman Cyril Q. Henriques wrote of the Negev: "With regard to its agricultural possibilities we can do no better than quote the Hope Simpson Report where it was stated that: 'Given the possibility of irrigation, there is practically an inexhaustible supply of cultivable land in the Beersheba area.'"[82] The temperatures in the Negev were described as a little different from those in other parts of the country. What worried experts more was the high wind velocity, a potential source of damage to crops. To solve this problem, as we have seen, Weitz suggested planting trees as windbreakers. Experts also were concerned by the levels of precipitation, which were significantly lower than in other parts of the country.

Nevertheless, since the Negev is characterized by seasonal floods, some suggestions were made to construct upper ground irrigation projects. The meteorologist Dov Ashbel believed, for instance, that dams should be built alongside *wadies* to capture their winter floods so that rainwater could be utilized for the irrigation of agricultural schemes in the area.[83] Another expert described the floods and their potential advantage for irrigation: "It appears that there are many *wadies* [near Beershbe] so formed by nature that artificial basins could easily be constructed of them to act as reservoirs, which would provide sufficient water for the irrigation of wide stretches."[84]

Another solution to transform the Negev into a productive territory was the utilization of dew for irrigation, which was presented as a pioneering Zionist invention. In a symposium on the agricultural potential of the Negev in June 1944, Ashbel explained that 1 millimeter of dew equaled 1 liter of water per dunam. In addition, in his observations he discovered that dew normally fell from 5:00 p.m. until 8:00 a.m. the next day, which meant that plants were covered with water for approximately sixteen hours a day. Ashbel claimed that this was much more significant than the availability of local precipitation. Like many other aspects examined in this study, the use of dew for irrigation was already practiced by Arab and Bedouin *fellahin* as a dry farming technique in the region. Dew mostly provided irrigation for the cultivation of watermelons and sorghum, which grew in summer.[85]

One of the most important Jewish experts who aimed to advance the utilization of dew in Palestine was Shimon Duvdevani, a Jewish biologist and teacher at the agricultural secondary school of Pardes Hanna. Between the years 1936 and 1943, Duvdevani established no few than eighty dew stations in Palestine, and his scientific work on the subject received growing attention among British experts across the empire, demonstrating once again the reciprocal exchange of knowledge between Zionists and other colonial enterprises, as well as the appropriation of Arab knowledge, which was made valuable for other colonizers via its Zionist credentialization.[86]

The use of dew was, of course, presented by Zionists and others as mirroring ancient biblical knowledge. For instance, an article from July 1947 in the British journal *Weather*, describing the work of Duvdevani, claimed that the use of dew in agriculture was an ancient Jewish practice, supported by old Hebrew prayers for summer dew while alluding to biblical and Talmudic references to dew as a useful source of irrigation for plants in Palestine.[87]

While dew was a local dry farming technique, the Zionist extraction of upper and ground water via advanced technology became, ironically, a tool with which to negate both the climate and the local vegetation. As we shall see in the last section of this chapter, Jewish agricultural experts not only often ignored the fact that Palestine was located within the Northern Hemisphere's desert belt, but also did not always aim to cultivate the plants and crops that grew in the region, choosing instead to introduce foreign, and potentially more profitable, crops.

FRUIT PLANTATIONS: A CASE STUDY IN ZIONIST ACCLIMATIZATION

Caroline Elkins and Susan Pedersen remind us that "pure" settler colonies, whose economy relied mostly on the labor and resources of the settler community and its patrons, quickly became integrated in global markets for crucial metropolitan goods. This was partly a result of their dislocation from the economies of the regions they inhabited, as well as their understanding of, and strong connections to, markets in their countries of origin.[88] As the cases of olive oil, sesame seeds, and citrus fruits exemplify,

Palestinian producers had been exporting goods to the international market since the eighteenth and nineteenth centuries when the coastal plain and its ports were strengthened by the Ottoman authorities. What was new and different about the Zionist pursuit of foreign markets was the attempt to introduce a variety of "foreign" fruits that required large quantities of water and thus technological solutions to overcome the local natural conditions. In order to address and identify the crops that would be best suited for such an export economy, Zionists first had to conduct a variety of climatic-agricultural experimentations in Palestine.[89]

As mentioned earlier, Zionist agronomists and farmers sometimes studied the cultivation of local crops, appropriating them with reference to their biblical origins. At other times they rejected these methods and crops altogether. However, this rejection was not just ideological: practical judgement likewise informed decisions relating to new Jewish plantations. For instance, some local crops required extensive time and effort to become fruitful and were not therefore seen by Zionists as economically viable. The most obvious example of this is the Zionist rejection of olive cultivation. Olive trees require an average of eight years before they begin to fruit, and twenty years to reach full production levels. Local olive trees, moreover, were known to have cycles of one good and one bad year.[90]

As mentioned earlier, the first grove (or forest, as it was called by contemporaries) initiated by the JNF in Palestine was meant to be composed of olive trees. But when Zionist experts realized, among other things, that olive trees required many years to grow and become fruitful, they decided to abandon their original plan. This approach can also be seen in the comments of the agronomist Elhanan Hershkowitz, who wrote many years later: "We could not compare our lifestyle and living standards to theirs [Palestinian Arabs] as well as to their patience towards trees and their willingness to grow trees that translate into an income only many years after being planted."[91]

The Jewish farmer and agronomist Shmuel Stoller also wrote, in a similar context, that when Zionists were interested in cultivating olive trees, they hoped to increase their productiveness by transforming them into an irrigated crop (olive trees were traditionally cultivated by dry farming). However, because the local olive tree, in fact, did not require large amounts of water, Zionist experts decided to import olive species from California, which required more irrigation. Stoller wrote that when olive cultivation indeed succeeded in Jewish plantations, it was due to this specie. In other words, by relying on "advanced" technology and knowledge concerning plant acclimatization, Zionist agronomists created a flawed agricultural logic that failed to comply with the local climate and environment.[92]

The decision to cultivate "foreign" crops was, first and foremost, determined by the Zionist material and technological advantage, which led Zionists to prefer agricultural schemes that would rely on modern irrigation solutions. As we shall see in the following pages, water proved a decisive element in establishing the type of crops planted by Zionists in Palestine, as well as the financial profit that could be made from them, and the national economy and nutrition that they would dictate.

The different climatic regions in the country, so often celebrated by Zionist experts as part of the unique characteristics of Palestine, also influenced agronomists' schemes for fruit growing. For instance, European apples, pears, peaches, cherries, and plums were planted in the cool mountain area, while in the humid regions of the country, Jewish farmers acclimatized tropical and subtropical fruits.[93] This endeavor succeeded solely because of artificial irrigation. In other words, although the temperatures of different climatic regions in the country could resemble those of "European" or "tropical" climates, precipitation usually did not match these climatic categories and therefore needed to be compensated for by artificial tools.

In *Cultivation of Sub-Tropical and Tropical Fruit Trees*, published in 1942, the botanist Hanan Oppenheimer discussed at length the foreign species that had been successfully acclimatized in Palestine in previous decades as a result of extensive irrigation. Among these fruits were mango, avocado, carambola, papaya, persimmon, loquat, lychee, guava, banana, pecan, and edible grapes.[94] Indeed, grapes had been traditionally cultivated in the country for centuries, but only with large quantities of water could the edible species of this fruit be cultivated without shrinking into raisins. Oranges, meanwhile, were so central to local agriculture (both Jewish and Arab) that the quantities of water they required[95] were often used as an indication for the cultivation of other fruits. Mango, for example, was reported as requiring 25 percent more water than citrus and avocado 50 percent more. Arabica coffee beans were also acclimatized in the experimentation station in Rehovot starting from the 1940s. This fruit required 1,875 millimeters of annual rain, which was two to four times the annual precipitation in Rehovot.[96]

Like grapes, bananas were reported to have been grown sporadically within Arab plantations along the coast and in the Jordan valley for centuries. However, the local species known as *musa acuminata* or *musa balbisiana* did not seem to be very popular among Jewish settlers. The most common banana species consumed in Western countries from the late nineteen century onwards was the Cavendish banana, which originated in the Canary Islands. Zionist agronomists considered this specific species the most suitable for the local climate, because it tolerated occasional cold weather and strong winds.[97]

The first Jewish banana plantation in Palestine was planted in Rothschild's estates in 1907, its stock imported from Algeria. In 1923, the Jewish communal settlement Kvuzat Kineret also began experimenting with different species of banana near the Sea of Galilee.[98] Stoller, who participated in these experiments, described in his memoir the textual sources that taught him how to cultivate this fruit (with which he was not previously familiar). These texts included a book dedicated to tropical and subtropical fruits, written by a German expert, and another book that focused specifically on banana cultivation, written by an English expert. As this example once again demonstrates, the knowledge absorbed by Jewish experts relied on European experience often accumulated in the colonies.[99] From the 1940s onwards, Jewish farmers also learned about the cultivation of tropical fruits in courses provided by the Union for Fruit Growers (*Irgun Megadlei Ha-Peirot*) which, among other objectives, encouraged local agricultural settlements to establish libraries that were supposed to include professional literature on the local climate and the cultivation of crops and be accessible to all Jewish farmers.[100]

Most banana plantations in Palestine were located on the coastal plain, as well as in the Beit Shean Valley, Jericho, and the Jordan Valley. These locations (especially the last three) were suitable for this purpose since according to contemporaries, they had a tropical climate and were close to underground water sources. Nonetheless, while bananas acclimatized successfully in these regions with the help of irrigation, Jewish laborers were reported to be experiencing difficulties when executing manual work

Figure 18: Irrigation in banana plantations in Mikve Israel, 1939.
Courtesy of the Kluger family and the Israel State Archives.

in these plantations.[101] A leaflet published by the Agriculture and Forestry Department of the British government in 1927 stated that Jewish settlers were not always interested in working or investing in banana plantations in the Jordan Valley: :If only they would have changed their work hours and wore hats . . . to protect themselves from the sun, the natural conditions [of the Jordan Valley] could be tolerable to them. Nevertheless, they must remember that they will not be able to expect a tropical yield crop without a tropical climate."[102]

The cultivation of bananas required an extensive amount of water. So great were the quantities of water required, in fact, that some plantations were harmed by extensive irrigation, and the abundance of water sometimes created a health hazard.[103] As a British report on a plantation in Jericho stated: "In the banana gardens near the spring there is a constant source of mosquitos. The best evidence of this is found in the fact that those people living near the garden have a higher percentage of parasites in their blood than those living in the more distant parts."[104]

While the terrace irrigation system was popular in other banana growing countries, in Palestine this method was not considered sufficiently advanced by Jewish farmers. Nevertheless, the intensive irrigation required by fruit industries sometimes threatened the general availability of local water.[105] A governmental memorandum from 1947 expressed "fear of [a] general lowering of the water table, following the extremely rapid development of orange cultivation. . . . The report further warns that lowering of the water table has given cause for concern."[106] Davis reminds us that these problems were prevalent in many arid environments where irrigation "without adequate drainage often results in salinization, waterlogging, the consequent loss of agricultural productivity, the spread of water-associated diseases, and other related environmental effects."[107] Indeed, according to Alon Tal, Jewish agriculture during the first half of the twentieth century caused increased salinity levels in groundwater as well as water contamination, and an overall depletion of local water sources.[108]

As we have seen, another flourishing subtropical crop that required large amounts of water was the legendary citrus, in particular oranges, which had thrived in Palestine since the mid-nineteenth century. For many decades the rainfall during the rainy season sufficed in Arab orchards. In the five dry months between May and October, most orchards in Palestine (except the few located near rivers) were irrigated by wells. For this reason, Arab orchards were planted in the coastal area where ground water was easy to access.

Following the introduction of deep drilling technologies by Western investors at the turn of the century, Jewish citrus growers soon expanded to locations beyond the coastal line. Arab growers were similarly taking the capitalist route in the citrus industry, and actually produced more citrus than Jews until the 1930s. However, it seems that this advantage was related to the greater use of manpower, and better locations near ground water and ports. In other words, Jewish orange growers were eventually able to compete with Arab growers mainly owing to the development of Western irrigation technologies, which negated the need for a large workforce and proximity to water sources. Indeed, advanced methods of irrigation increased citrus exports from fewer than one million cases in 1921 to more than 15 million cases in 1939, representing 74 percent of the total agricultural export produce of the country.[109]

California and its large irrigation projects provided once again an important role model for Jewish citrus growers. Stoller wrote, for instance, that when aiming to cure the unsuccessful orange grove in Kineret, he tried to copy the image of a healthy grove that he saw in an American agriculture journal depicting citriculture in California.[110] As Karlinsky reminds us, the Californian model was followed by Jewish citrus growers not only because of the climatic resemblance of these countries, but also because of the availability of private capital, which enabled large-scale artificial irrigation.[111] Warburg had already commented on this very issue in 1912: "Above all, we are able, by virtue of our propaganda, to introduce into the country people with capital; for with the help of capital, as California shows, garden cultivation can develop enormously in a few years, if the country and the population are fitted for it."[112]

As mentioned above, profit from citriculture depended mainly on export.[113] During most of the nineteenth century, oranges as well as bananas were luxury items in the West. They were expensive, and most people had never seen them before. However, during the early twentieth century, this changed with the growth in real incomes that followed increasing industrialization. Specifically, during the interwar period, bananas and oranges became the most popular imported fruits in Britain. As a result, the position of Palestine as a fruit exporting country increased immensely under Mandate rule. Between the years 1927 and 1931, Palestine became the fourth-largest exporter of citrus in the world, and by 1941 it was ranked second after Spain.[114]

Various local and international reports demonstrated the importance of agriculture for exports, and one article from the Palestine Post even declared that British food experts in the post-World War II era expected Palestine to become a "great food growing country" and an aid to "hungry Europe."[115] Other foreign experts claimed that although the climate of Palestine was not perfect for growing bananas, the country's location was perfect for the European markets. They wrote that this was especially true for Eastern Europe where hitherto goods from places such as the Canary Islands or Central America "had to go through many hands" before arriving safely at their final destination.[116] Thus, in the international fruit market in particular, water was seen as a means to expand commercial agriculture through productive control of nature, which was meant to increase revenues.

Zionist experts had expressed similar hopes from the 1920s, relying on the promise of irrigation technologies. According to Zelig Soskin, irrigation could double, and even triple, annual levels of local production. He explained that countries with subtropical climates, such as Palestine, if artificially irrigated, have an advantage over temperate countries as they are blessed with intense sunlight and mild winters. Soskin believed that "the extension of the process of vegetation by means of an artificial supply of water also during the dry season would make possible a second, and even a third and fourth, crop."[117]

Yet, while the orange export market was booming throughout the first half of the century, bananas and other tropical and subtropical fruits were still being tested and discussed and did not stabilize as export commodities until the late 1940s. This was partly the result of the further need for Jewish experts to study the specific conditions of different species in different climatic regions within the country. Moreover, the fact that many fruit trees required several years until they bore profitable crops did not help the realization of such economic ambitions. Finally, knowledge on the best ways to transport fruits in long distances also had to be accumulated.[118]

Climate investigation was vital to the fruit market, as it indicated the desired natural conditions for its successful cultivation. This knowledge also came in handy when aiming to expand the fruit supply throughout the year in order to increase local profitability. Hershkowitz wrote in this context:

> Our country is indeed small, but it is blessed with diverse regions and different combinations of climate and soil. . . . After planning plantation according to different regions and following a thorough study of each region, including its topographical, geological and climatic conditions as well as investigating the blooming and ripening period of each region—I arrived at the conclusion that there are endless possibilities for the expansion of plantations for a regular supply of fruits in quantities that will suffice the consumption of the Jewish community and even create surplus for the industry and for export.[119]

In this text, Hershkowitz explained how different sorts of edible grapes, cherries, peaches, and plums (which were grown in different climatic regions of the country and therefore ripened at slightly different times of the year) could be sold to the market from May until November—that is, for 170 days instead of the usual one or two months in which fruit grown and harvested in one climatic region could be sold.[120]

However, despite the economic ambitions that were intertwined with the cultivation of fruits, Jewish settlers in Palestine often had to be encouraged to consume them (especially tropical fruits, which they viewed as strange and unfamiliar). The agronomist Akiva Etinger wrote in a magazine for housewives that:

> During half of a year and longer, we could use the abundance of fruits which are produced by the citriculture industry. . . . Do we use this wonderful abundance enough? Not at all. The housewife disregards these excellent and cheap fruits, which cost a fortune in a few countries around the world where they cannot be grown while she prefers—without any justification—fruits that are brought to us from foreign countries [Europe] and are very expensive.[121]

A similar statement was made by Professor D. Rosenbaum, who wrote in 1939: "Our mothers feed their children on [imported] apples and even pears, which are still more expensive, without considering the fact that the vitamin C in apples is less than one-third that of oranges, while the pear hardly contains any vitamins at all. We are in the ridiculous position of spending good money on substitutes despite an over-abundance of excellent fruit."[122]

Such recommendations followed the same lines as, and were indeed the continuation of, the nutritional advice presented in chapter 3. However, from 1936 they also had clear national and political connotations. The outbreak of the Palestinian Revolt in 1936 not only harmed the economic relationships between Jews and Arabs in the country. It also narrowed the access of the Jewish community to local Arab produce. Moreover, Arab port strikes significantly increased prices for imported goods. These circumstances further convinced Zionist leaders of the need to create a separate and independent Jewish economy. As part of this realization, they pressed for increased Jewish agricultural production and greater segregation between the two local economies. This initiative was also related to the JA's establishment of the Union for the Land's Produce (*Totzeret Ha-Aretz*) in 1936, an initiative that promoted Jewish produce. Thus, for Zionists the cultivation of fruits through extensive irrigation systems meant an increase in local capital by export, while simultaneously providing work for as many Jewish people as possible.[123]

Although agriculture and forestry are often associated in historiography with the Zionist ideology that advocated the transformation of an allegedly weak, pallid, and effeminate populace into a strong, tanned, and masculine one, these activities were also meant to transform and "cure" the land itself, improve its climate, enhance its hygiene, and establish the settler society's economy. As we have seen, climate was a key element in understanding and developing Zionist agriculture and forestry in Palestine. As a result of its potential influence on the success of food production in the country, the investigation of climate was also intertwined with other fundamental aspects, such as water management, the national economy, and global food markets. Attempts to manage local climate led to its being compared to other, more "familiar" geographies, and simultaneously inspired imaginative reconstructions of Palestine's climate in biblical times. Furthermore, efforts to negate and improve climate were associated with Zionist national ideology, and eventually had a significant material impact on the Arab-Jewish conflict.

Conclusions

Today we are faced with the perils of climate change, which is already witnessed in many parts of the world, including in the Middle East and specifically in Israel/Palestine. The implications of rising temperatures and extreme weather events have been shown to not only affect people's thermal comfort but also to influence their immunity to disease and their agricultural production, thus causing, among other things, disruptions to global food trade and subsequent food shortages.[1]

In this context, this book was not easy to write. Although I point a critical finger towards the evolution of Zionist climate investigation in Palestine and its colonial and Eurocentric biases and premises, I also acknowledge and fully comprehend the very real climatic perils faced by the Middle East (including Israel/Palestine) today. Moreover, despite the primary focus of this study on the history of Jewish orientalist environmentalism, while hardly addressing the issue of climate change itself, I believe that this theme is, nevertheless, pertinent for our understanding of some of the most serious issues of our time in this region.

The rise in wildfires, decline in biodiversity, scarcity of water, and changes in soil composition that are experienced in today's Israel/Palestine are largely a result of the scientific and technological interventions introduced to this country by modern colonial, capitalist, and orientalist forces. These changes are also a result of specific Zionist actions that were meant to "fight" the local climate, including the decisions to transform the landscape and ignore its natural ecosystems, to plant monocultural forests and cash crop plantations, to irrigate them intensively, and to use polluting technologies (including, among other things, air conditioning).[2]

By focusing on the Zionist colonial and orientalist approach to Palestine's climate, environment and local population, this study also stressed the arrogance that often characterized this approach which persisted in Israeli culture throughout the following decades. For example, in 1981 the columnist and artist Amos Kenan wrote in *Your Land, Your Country*:

> The summer is the season in which we do not know how to live in this country. It is too hot for us. People usually learn how to live in their country. . . . We brought here very much but took so little from this place. There is arrogance not only in our attitude towards the Arabs who surround us—we are always confident that we have something to teach them and not so much to learn from them—we are also arrogant towards our country. We dressed her in a gown of concrete and cement . . . as the poem says, but we did not ask the hot desert winds how to coexist with them, we did not ask the burning sun what it can offer us, we did not ask the rocks and the soil, perhaps they have an answer.[3]

Kenan's poetic depiction of the Israeli sense of superiority over the environment and local forms of environmental knowledge explains how, despite a growing international concern about the impacts of climate change today (especially in the Middle East), some Israeli scientists remain unscathed. For example, Professor Dani Rosenfeld, a climate scientist at the Hebrew University of Jerusalem, claims that "in Israel, we have a better chance of fighting off many of the problems that will develop [as a result of climate change] due to our use of desalinization, irrigation, [and] air-conditioning. . . . This can surely give the citizens a better feeling."[4] Such statements also reflect the prevailing belief in the power of technology to fight and triumph over nature that has been one of the central subjects of investigation in this book.

Moreover, by describing the orientalist and colonial approach of Jewish scientists to the local climate and the environment, this study demonstrated the early days of the ongoing relationship between environmental and human injustices in Palestine that some scholars have recently termed "eco-apartheid."[5] One example of this relationship is evident in the activity of the JNF, which proudly declares itself today as the biggest and most important ecological entity in Israel, and one of the leading forces in "fighting off" the desert as a result of tree planting in the Negev. The Yatir Forest, planted in this region by the JNF in the 1960s and composed mostly of pine trees, is not only the biggest forest in Israel today, but also the biggest human-made forest in a semi-arid environment in the world. Such actions are controversial as they ignore recent research that has shown that afforestation projects in arid regions during the nineteenth and twentieth centuries have often failed, or even harmed the natural environment. Davis writes in this context that "where reforestation/afforestation projects have 'succeeded' in terms of the trees surviving, they have frequently used so much groundwater that local water tables have been lowered, wells have run dry, and nearby soils have been desiccated, reducing agricultural yields."[6]

The Yatir Forest has also received attention in recent decades because it has become a central arena in which the State of Israel is fighting its Bedouin citizens. Bedouin villages such as Hassein Al Rafiaa and Al-Araqeeb have faced forced removal as a result of afforestation schemes since the late 1990s, despite there being plenty of other territories available in the Negev in which to pursue these projects. Unfortunately, Israel's contemporary attempts to "delay desertification processes" not only reflect a contested and potentially harmful act towards the local arid ecosystem. They also mirror ethnocratic territorial aspirations.

As we have seen in this study, the orientalist Zionist approach to Palestine's environment and its local population did not imply a simplistic and one-dimensional relationship. On the contrary, it reflected a nuanced spectrum of approaches to both the newly acquired territory and its natural characteristics, as well as its indigenous people. These approaches were influenced by both metropolitan knowledge, knowledge produced and accumulated by Westerners in other colonial and settler colonial contexts, as well as local Palestinian knowledge. On the one hand, the climate and environment in Palestine were compared with the Central European environments that the settlers were forced to leave behind, yet for which they still longed. On the other hand, the local climate was often compared with other settler territories such as California, Queensland, and French Algeria from which Jewish settlers aspired to learn how to overcome the hazards of warm climates. Finally, Zionists often also studied local methods of dealing with the climate in Palestine and appropriated them by stressing their supposed ancient biblical roots.

This multiple use of knowledge sources and references to landscapes can be explained by one of the most salient traits of the Zionist project: the liminality of its members between East and West. While Jewish Europeans sometimes wished to view themselves as Occidentals settling in the Orient, at other times, they hoped to take root in the country and become native to it. Thus, while like other Europeans they feared the consequences of warm climates on their bodies, Jewish settlers were often also open to studying and at times even copying Palestinian methods of managing with the local climate (even while simultaneously adhering to technological solutions developed in Western and other settler countries).

Another related theme that was highlighted in this book concerns the simultaneous rejection and absorption of indigenous Arab methods and knowledge by Zionists. From the first decade of the twentieth century, Jewish settlers began demonstrating settler-colonial characteristics in the rhetoric and practices of their national project. They usually did not wish to govern the local Arab population or to recruit it in their economic undertakings. Instead, they were more interested in acquiring the Arabs' land and hoped to eliminate their presence. Nevertheless, before 1948, the elimination of the indigenous presence in Palestine was not necessarily expressed in physical violence or annihilation, but rather in the form of segregation and appropriation—two actions that might initially appear contradictory.

During the period discussed in this book, the segregation of Jews and Arabs was both formal and informal, and was exercised in the political, cultural, economic, and physical spheres. This approach also pertained to processes of knowledge production concerning the local climate and environment and was often linked with feelings of anxiety about, and disgust at, the local population and their lifestyle, as it tended to associate the local climate and environment with so-called Palestinian neglect.

Jewish appropriation of local Arab customs and knowledge pertaining to climate also contributed to indigenous elimination. According to Rayna Green, the cultural appropriation of indigeneity is based on a logic of genocide in which non-native peoples imagine themselves as the rightful inheritors of what previously belonged to the local population, thus entitling them to ownership of the land.[7] This kind of appropriation was manifested in the Zionist context through its harnessing of the Hebrew Bible, which was often used by experts to justify the adoption of indigenous solutions for coping with the local climate.

At other times, the Bible was used as a so-called reminder of the past existence of a 'better' climate, which had deteriorated as a result of local neglect, and which needed to be restored by the colonizers. Of course, biblical sentiments existed as foundational myths in the development of many national movements, especially settler-colonial movements. However, while other nations used the Bible and its names and places in a metaphoric fashion, Zionists aimed to use them in a literal way.

In this study, I have focused on the fields of expertise that were most affected by, and occupied with, the issue of climate. As we have seen, these fields included climatology and meteorology, race sciences and medicine, architecture and planning, and agriculture and botany. Despite the exchange of knowledge between experts in these fields, each scientific discipline operated within specific strictures and harbored specific concerns. In addition, each of these fields evoked general popular reactions to, and perceptions of, climate among European Jewish settlers, thus enabling me to occasionally discuss in tandem the cultural construction of expert knowledge.

The 1948 Arab-Israeli War and the ensuing establishment of the State of Israel changed the geopolitics of the region, and immediately transformed many aspects of the politics, economy, society, and culture of Palestine's Arab and Jewish communities. As Kenan wrote several decades later, these changes did not necessarily influence the ways in which climate in the country was understood and contended with by the Jewish-Israeli society. On the contrary, from the 1960s Israel became globally famous for its national triumph over nature. This triumph has been manifested mainly in the development of advanced irrigation technologies, which have subsequently been exported to other countries with warm and dry climates. During the following decades, Israel also developed a pioneering air-conditioning industry, and its climatological studies in architecture contributed significantly to the global development of new building methods.

An interesting subject for further research on this topic would be the case study of the Negev, in relation to the climatic "challenges" it evoked in Zionist eyes after 1948. The Negev is a large desert area, which comprises about half of pre-1967 Israel. As we have seen, the position of the Negev in pre-statehood Zionist thought was relatively unclear and undetermined, and only a very few Jewish settlers were willing to dwell there during the British Mandate. However, in 1947 the United Nations partition plan determined that most of the Negev would become part of the future Jewish state. This decision transformed the status of the Negev and the meanings attributed to it overnight and increased official statements on the need to "redeem" it from its desolate conditions. Aspirations to paint the Negev green were embedded in a political discourse of anxiety regarding the "emptiness" of this contested territory and the urgent need to settle it. The reconfiguration of the Negev, the development of innovative irrigation technologies to meet its unique characteristics, and the experimental agricultural projects initiated there during the 1950s and 1960s (which included at some point the development of artificial climate control to increase rainfall) demonstrated how Western science and technology were used to overcome nature itself.[8]

Additional historical and sociological research is also needed to examine why contemporary discourses on climate change have taken little root in Israel today. In Western countries, to which the Israeli public and authorities continue to compare themselves, the discourse on climate change is often associated with other discourses on science, politics, economy, religion, and ethics. It appears that the absence of a similar discourse in the Jewish-Israeli political and cultural sphere is not disconnected from the colonial history of this country, or the relationship of the Jewish society to the local climate, both of which I have discussed in this study. The local political, cultural, and religious ambiguity embedded in this relation, on the one hand, and the Zionist aim to negate the longstanding Jewish reality of rootlessness and displacement, on the other, appear to hinder the Israeli absorption of new political ideas on the global dimensions of human accountability for climate change.

I opened this concluding chapter by explaining why this book was not easy to write at a time of increasing global and local environmental concerns and crises. A moment before this book is being published, I face another great challenge. The Israel–Hamas War that broke out on October 7, 2023 is not only a difficult moment for all those who stand with humanity. I believe it is also a challenging moment for those who work on colonial and postcolonial studies and who are now required to reflect on their premises and arguments.

There should be no denial of the fact that settler colonials, among them Zionists, have often violently occupied territories of other people and wished to replace them in order to become indigenous themselves. Nevertheless, I believe that a more nuanced application of the settler-colonial framework to Zionist history must be taken. It should be remembered that settler colonials were often also moved by force of circumstances or necessity as migrants, convicts, refugees, and seekers of wealth. As I have demonstrated throughout this book, this was especially true in the case of Jewish settlers in Palestine, who were persecuted in Europe, among other things, as a result of ethno-racial separatist processes. Indeed, it seems like the origin of settler colonial movements has not always received enough scholarly attention within this analytical framework. This is not to imply that the political and economic reasons that pushed settlers to new destinations make their projects just. However, these reasons do illuminate settler societies in more complex ways and, moreover, they should remind us, especially in relation to the Israeli-Palestinian conflict, that simplistic calls for the "decolonization" of Palestine "from the river to the sea" are neither useful nor feasible. A complex historical condition requires a complex contemporary solution, which I hope will be found sooner rather than later for this anguished and bloody conflict.

Notes

INTRODUCTION

1. Bruno Latour, *We Have Never Been Modern* (Cambridge, MA: Harvard University Press, 2011), 11. Italics original.

2. The Jewish-dominated, rather than a comparative or relational Jewish-Arab, framework is a result of two main factors. The first reason is related to the historical evolution of colonial science in general, and of climate science in particular, as will be explained in the following pages. The second reason is that my own biographical and educational background has partly determined my access to certain sources and has led me to ask specific questions. I hope other scholars who possess other types of backgrounds and education will contribute to this study by offering a more in-depth Palestinian perspective on scientific and popular approaches to the local climate.

3. Edward W. Said, *Orientalism* (New York: Penguin Books, 1995 [1978]), 2.

4. Eitan Bar-Yosef, *The Holy Land in English Culture 1799–1917: Palestine and the Question of Orientalism* (Oxford: Clarendon, 2005), 8; Said, *Orientalism*, 34.

5. Diana K. Davis and Edmund Burke III, eds., *Environmental Imaginaries of the Middle East and North Africa* (Athens: Ohio University Press, 2011), 16.

6. As a result of this discourse, recent studies of the Middle East have adopted the perspectives of history of science and technology, as well as environmental history, to discuss colonialism and the image of the East through Western eyes.

These studies include Alan Mikhail, *Nature and Empire in Ottoman Egypt : An Environmental History* (Cambridge, UK: Cambridge University Press, 2011); Timothy Mitchell, *Rule of Experts : Egypt, Techno-Politics, Modernity* (Berkeley: University of California Press, 2002); Robert Fletcher, *British Imperialism and 'the Tribal Question': Desert Administration and Nomadic Societies in the Middle East, 1919-1936* (Oxford: Oxford University Press, 2015); Onur Inal and Yavuz Kose, *Seeds of Power : Explorations in Ottoman Environmental History* (Winwick: The White Horse Press, 2019); On Barak, *Powering Empire : How Coal Made the Middle East and Sparked Global Carbonization* (Oakland: University of California Press, 2020).

7. Davis and Burke, *Environmental Imaginaries*, includes a chapter on Israel/ Palestine written by Shaul Cohen. However, this chapter fails to address the environmental history of this country from a postcolonial perspective.

8. See, for example, Richard Grove, *Green Imperialism: Colonial Expansion, Tropical Island Edens and the Origins of Environmentalism, 1600-1860* (Cambridge, UK: Cambridge University Press, 1995); Mark Harrison, *Climates & Constitutions: Health, Race, Environment and British Imperialism in India, 1600-1850* (Oxford: Oxford University Press, 1999); David Arnold, *Warm Climates and Western Medicine: The Emergence of Tropical Medicine, 1500-1900* (Amsterdam: Rodopi, 1996); Eric T. Jennings, *Curing the Colonizers: Hydrotherapy, Climatology, and French Colonial Spas* (Durham: Duke University Press, 2006); Diana K. Davis, *The Arid Lands: History, Power, Knowledge* (Cambridge, MA: The MIT Press, 2016). Michael Osborne, *The Emergence of Tropical Medicine in France* (Chicago: University of Chicago Press, 2014); Warwick Anderson, *The Cultivation of Whiteness: Science, Health and Racial Destiny in Australia* (Carlton, Australia: Melbourne University Press, 2005).

9. Grove, *Green Imperialism*, 14, 76; Londa Schiebinger and Claudia Swan, eds., *Colonial Botany: Science, Commerce, and Politics in the Early Modern World* (Philadelphia: University of Pennsylvania Press, 2005), 2-7.

10. Deborah R. Coen, "Climate and Circulation in Imperial Austria," *The Journal of Modern History* 82, no. 4 (December 2010): 839-75; Martin Mahony, "For an Empire of 'All Types of Climates': Meteorology as an Imperial Science," *Journal for Historical Geography* 51 (2016): 29-39.

11. Manuscript by Rudolf Feige, unknown date, AGSJI, G.F.0341-4/2.

12. For example, Lorenzo Veracini, *Israel and Settler Society* (London: Pluto, 2006); Patrick Wolfe, "Settler Colonialism and the Elimination of the Native," *Journal of Genocide Research* 8, no. 4 (2006): 387-409; Gershon Shafir, *Land, Labor and the Origins of the Israeli-Palestinian Conflict 1882-1914* (Cambridge, UK: Cambridge University Press, 1989); Areej Sabbagh-Khoury, "Tracing Settler Colonialism: A Genealogy of a Paradigm in the Sociology of Knowledge Production in Israel," *Politics and Society* 50, no. 1 (2022): 44-83; Derek Jonathan Penslar, *Zionism and Technocracy: The Engineering of Jewish Settlement in*

Palestine, 1870–1918 (Bloomington: Indiana University Press, 1991); Rashid Khalidi, *The Hundred Years' War on Palestine: A History of Settler Colonial Conquest and Resistance* (London: Profile Books, 2020); Shira Robinson, *Citizen Strangers : Palestinians and the Birth of Israel's Liberal Settler State* (Stanford, CA: Stanford University Press, 2013); Gabriel Piterberg, *The Returns of Zionism: Myths, Politics and Scholarship in Israel* (London: Verso, 2008); Rachel Busbridge, "Israel-Palestine and the Settler Colonial 'Turn': From Interpretation to Decolonization," *Theory, Culture & Society* 35, no. 1 (2018): 91–115.

13. For example, Tamar Novick, *Milk and Honey: Technologies of Plenty in the Making of a Holy Land* (Cambridge, MA: MIT Press, 2023); Irus Braverman, *Settling Nature: The Conservation Regieme in Palestine-Israel* (Minneapolis: University of Minnesota Press, 2023); Irus Braverman, *Planted Flags: Trees, Land, and Law in Israel/Palestine* (Cambridge, UK: Cambridge University Press, 2009); Samer Alatout, "Bringing Abundance into Environmental Politics: Constructing a Zionist Network of Water Abundance, Immigration, and Colonization," *Social Studies of Science* 39, no. 3 (2009): 363–94; Matan Kaminer, "Towards a Political Ecology of Zionism in the Rural Sphere," *Teorya U-Viḳoret (Theory and Criticism)* 57 (Winter 2023): 71–99; Mazin B. Qumsiyeh and Mohammed A. Abusarhan, "An Environmental Nakba: The Palestinian Environment under Israeli Colonization," *Science Under Occupation* 23, no. 1 (Spring 2020); Efrat Gilad, "Camel Controversies and Pork Politics in British Mandate Palestine," Global Food History (2022): https://doi.org/10.1080/20549547.2022.2106074.

14. Braverman, *Settling Nature,* 7.

15. Derek Jonathan Penslar, "Zionism, Colonialism and Technocracy: Otto Warburg and the Commission for the Exploration of Palestine, 1903–7," *Journal of Contemporary History* 25, no. 1 (January 1990): 146–47; Frank Leimkugel, *Botanischer Zionismus: Otto Warburg (1859–1938) Und Die Anfänge Institutionalisierter Naturwissenschaften in 'Erez Israel'* (Berlin: Veröffentlichungen aus dem Botanischen Garten und Botanischen Museum Berlin-Dahlem, 2005), 3, 29, 31; Dana Von Suffrin, "The Possibility of a Productive Palestine: Otto Warburg and Botanical Zionism," *Israel Studies* 26, no. 2 (Summer 2021): 173–97.

16. Penslar, *Zionism and Technocracy,* 61.

17. Robinson, *Citizen Strangers,* 2.

18. Bar-Yosef, *The Holy Land in English Culture*; Barbara W. Tuchman, *Bible and Sword : England and Palestine from the Bronze Age to Balfour* (London: Papermac, 1982).

19. The British interest in Palestine was, moreover, related to the geopolitical and economic possibilities of this country. Palestine's location between east and west linked three continents and starting from the late nineteenth century was considered as an area necessary for the defense of the Suez Canal, the road to India, and the oil fields of Mosul; see Tuchman, *Bible and Sword,* viii.

20. Bar-Yosef, *The Holy Land in English Culture*, 8–9.

21. When making the distinction between colonialism and settler colonialism, it is important to remember that scientific approaches to warm climates cannot always be classified neatly between these two categories, as they usually developed in tandem despite reflecting different colonial circumstances.

22. Eliezer Livneh, *Aaron Aaronsohn: Ha-Ish Ve-Zmano* (Jerusalem: Bialik, 1969), 119–43.

23. Aaron Aaronsohn, *Agricultural and Botanical Explorations in Palestine* (U.S. Department of Agriculture, Bureau of Plant Industry, Bulletin N180, 1910), 17–30. Aaronsohn opposed the separatism advocated by settlers of the second immigration wave (1904–14) as part of their campaign for "Hebrew Labor," making him a rather complex figure.

24. Caroline Elkins and Susan Pedersen, *Settler Colonialism in the Twentieth Century: Projects, Practices, Legacies* (London: Routledge, 2005), 2; Tracey Banivanua-Mar and Penelope Edmonds, *Making Settler Colonial Space: Perspectives on Race, Place and Identity* (Basingstoke, UK: Palgrave Macmillan, 2010), 2; Patrick Wolfe, "Land, Labor, and Difference: Elementary Structures of Race," *The American Historical Review* 106, no. 3 (2001): 866–905; Lorenzo Veracini, *Settler Colonialism: A Theoretical Overview* (New York: Palgrave Macmillan, 2010), 33–52.

25. Lorenzo Veracini, "The Other Shift: Settler Colonialism, Israel, and the Occupation," *Journal of Palestine Studies* 42, no. 2 (2013): 28. Unlike Veracini, who views separatist approaches to indigenous populations as characterizing colonial behavior and, thus, as contrary to settler colonial approaches of appropriation, I argue that the two approaches are not necessarily contradicting in settler societies that have not yet 'normalized'.

26. Rayna Green, "The Tribe Called Wannabee: Playing Indian in America and Europe," *Folklore* 99, no. 1 (1988), quoted in Andrea Smith, "Queer Theory and Native Studies: The Heteronormativity of Settler Colonialism," *GLQ* 16, no. 1-2 (2010): 53; Wolfe, "Land, Labor, and Difference." Other works that discuss settler colonial elimination through appropriation or assimilation in Israel/Palestine include: Robinson, *Citizen Strangers*; Arnon Degani, "On the Frontier of Integration: The Histadrut and the Palestinian Arab Citizens of Israel," *Middle Eastern Studies* 56, no. 3 (2020); Novick, *Milk and Honey*.

27. For example, Ori Yehudai, *Leaving Zion: Jewish Emigration from Palestine and Israel after World War 2* (Cambridge, UK: Cambridge University Press, 2020); Tara Zahra, *The Great Departure: Mass Migration from Eastern Europe and the Making of the Free World* (New York: W. W. Norton and Company, 2016); Gur Alroey, *An Unpromising Land: Jewish Migration to Palestine in the Early Twentieth Century* (Stanford, CA: Stanford University Press, 2014); Shafir, *Land, Labor and the Origins of the Israeli-Palestinian Conflict*.

28. More than 74 percent of the Jewish population during the Mandate period was urban, and the number of Jews working in the agricultural sector decreased between the years 1922 and 1945, from 27 percent to 13 percent. On the contrary, in the industry sector it grew from 17 percent to 31 percent. See Jacob Metzer, *Economics, Land and Nationalism: Issues in Economic History and Political Economy in the Mandate Era and the State of Israel* [Hebrew] (Jerusalem: Magnes Press, 2023), 17.

29. Hizky Shoham, *Carnival in Tel Aviv : Purim and the Celebration of Urban Zionism* (Boston, MA: Academic Studies Press, 2020); Anat Helman, *Young Tel Aviv : A Tale of Two Cities* (Hanover: University Press of New England, 2010), 91-92; Efrat Gilad, "Meat in the Heat: A History of Tel Aviv under the British Mandate for Palestine (1920s-1940s)" (PhD diss., University of Geneva, 2021), 11, 29-30.

30. The latter specifically took place during the 1930s (Metzer, *Economics, Land and Nationalism*, 14-15).

31. See, for example, Yael Zerubavel, *Desert in the Promised Land* (Stanford, CA: Stanford University Press, 2019); Maya Duany, *Ha'agam Ha'neelam: Hachula Khesemel Le'hishtanut Ha'yachas La'sviva Be'medinat Israel* (Jerusalem: Yad Ben-Zvi, 2020); Oz Almog, *The Sabra—A Profile* [Hebrew] (Tel Aviv: Am Oved, 1997); Alon Tal, *Pollution in a Promised Land: An Environmental History of Israel* (Berkeley: University of California Press, 2002).

32. These ideas are also evident in the following studies: Dafna Hirsch, *"We Are Here to Bring the West": Hygiene Education and Culture Building in the Jewish Society of Palestine During the British Mandate Period* [Hebrew] (Sde Boker: The Ben-Gurion Research Institute for the Study of Israel and Zionism, 2014); Ethan Katz, Lisa Moses Leff, and Maud Mandel, "Introduction: Engaging Colonial History and Jewish History," in *Colonialism and the Jews*, ed. Ethan Katz, Lisa Moses Leff, and Maud Mandel (Bloomington: Indiana University Press, 2017).

33. Billie Melman, *Women's Orients : English Women and the Middle East, 1718-1918 : Sexuality, Religion and Work*, 2nd ed. (Basingstoke, UK: Macmillan, 1995), 29; Novick, *Milk and Honey*, 8-9.

34. Anthony D. Smith, *Chosen Peoples* (Oxford: Oxford University Press, 2003), 77-78; Amnon Raz-Krakotzkin, "En Elohim, Aval Hu Hevtiach Lanu Et Ha-Aretz," *Mita'am* 3 (2005): 71-76. It should be stressed that nationalism does not need to contradict the concept of settler colonialism. As Gabriel Piterberg writes, for instance, Zionism "was both a Central-Eastern European national movement *and* a movement of European settlers which sought to carve out for itself a national patrimony with a colony in the East" (*The Returns of Zionism*, xii). For further reading on the link between nationalism and settler colonialism, see Shafir, *Land, Labor and the Origins of the Israeli-Palestinian Conflict*;

Khalidi, *The Hundred Years' War on Palestine*; Sabbagh-Khoury, "Tracing Settler Colonialism."

35. Raz-Krakotzkin, "En Elohim," 72.

36. Avner Ben-Amos and Ofri Ilany, *Sacred People: Bible and Nationalism in the Modern Era* [Hebrew] (Jerusalem: Magnes Press, 2021), 9, 10, 17.

37. Yaacov Shavit and Mordechai Eran, *The Hebrew Bible Reborn: From Holy Scripture to the Book of Books: A History of Biblical Culture and the Battles over the Bible in Modern Judaism* (Berlin: Walter de Gruyter, 2007). A new approach to the Bible is already notable in the writings of eighteenth-century *maskilim* who wished to replace "'the twisted rabbinic commentary of the scriptures' and 'to introduce a rational exegesis.'" Alternatively, the emphasis in the late nineteenth century was on the national role of this text. See Moshe Pelli, *Haskalah and Beyond : The Reception of the Hebrew Enlightenment and the Emergence of Haskalah Judaism* (Lanham, MD: University Press of America, 2010), 30–31.

38. Diana K. Davis, *Resurrecting the Granary of Rome: Environmental History and French Colonial Expansion in North Africa* (Athens: Ohio University Press, 2007), 58.

CHAPTER 1. KNOWING CLIMATE

1. Deborah Joy Neill, *Networks in Tropical Medicine: Internationalism, Colonialism, and the Rise of a Medical Specialty, 1890–1930* (Stanford, CA: Stanford University Press, 2012), 6–7.

2. Kristine Harper, *Weather by the Numbers: The Genesis of Modern Meteorology* (Cambridge, MA: MIT Press, 2008), 11; Frederik Nebeker, *Calculating the Weather: Meteorology in the 20th Century* (San Diego: Academic Press, 1995), 45; Philipp Lehmann, *Desert Edens: Colonial Climate Engineering in the Age of Anxiety* (Princeton: Princeton University Press, 2022), 3.

3. Hilel Yaffe and Soskin Selig, *Log Book of the Commission Appointed to Report on the Practicability of Establishing Settlments on the Land under Egyptian Administration East of the Suez Canal and Gulf* (ISA פ—19 / 1950, 1903), 14, 1.

4. The use of both temperature systems (F and C) was common in research expeditions, usually a result of the employment of different types of equipment.

5. Nebeker, *Calculating the Weather*, 15.

6. Ibid.

7. Yoffe and Selig, *Log Book of the Commission*, 29.

8. Mitchell Bryan Hart, *Social Science and the Politics of Modern Jewish Identity* (Stanford, CA: Stanford University Press, 2000), 5; Nebeker, *Calculating the Weather*, 21, 23.

9. Hart, *Social Science and the Politics of Modern Jewish Identity*, 3–6, 17.

10. Yoffe and Selig, *Log Book of the Commission*, b1–3.

11. Hilel Yaffe, *Pirkei Zikhronot* (Tel Aviv: n.p., 1935), 27. This claim is also based on the report of the irrigation engineer William Garstin of the Anglo-Egyptian administration. See Robert G. Weisbord, *African Zion: The Attempt to Establish a Jewish Colony in the East Africa Protectorate, 1903–1905* (Philadelphia, Jewish Publication Society of America, 1968), 57.

12. Gur Alroey, "Journey to New Palestine: The Zionist Expedition to East Africa and the Aftermath of the Uganda Debate," *Jewish Culture and History* 10, no. 1 (May, 2012): 29.

13. Quoted in Tara Zahra, "Zionism, Emigration, and East European Colonialism," in *Colonialism and the Jews*, ed. Ethan Katz, Lisa Moses Leff, and Maud Mandel (Bloomington: Indiana University Press, 2017), 175.

14. For more detail see Netta Cohen, "Shades of White: African Climate and Jewish European Bodies, 1903–1905," *The Journal of Imperial and Commonwealth History* 50, no. 2 (2022).

15. Before joining the Zionist expedition to Guas Ngishu, Kaiser's personal research had focused on precipitation in the Sinai Peninsula with the aim of proving the fertility of this region during ancient biblical times.

16. Eitan Bar-Yosef, "Spying out the Land: The Zionist Expedition to East Africa, 1905," in *'The Jew' in Late-Victorian and Edwardian Culture: Between the East End and East Africa*, ed. Eitan Bar-Yosef and Nadia Valman (Basingstoke, UK: Palgrave Macmillan, 2009), 190–92.

17. Major A. St. Hill Gibbons, Alfred Kaiser, and Nahum Wilbush, *Report on the Work of the Commission Sent out by the Zionist Organization to Examine the Territory Offered by H.M. Government to the Organization for the Purpose of a Jewish Settlement in British East Africa* (London: Wertheimer, Lea & Co, 1905), 11–12.

18. Ibid., 43.

19. Ibid., 57.

20. Ibid., 69.

21. Ibid., 81.

22. Ibid., 72.

23. Indeed, Wilbush was not a great supporter of the East Africa scheme. However, neither did he reject it. One pragmatic approach of the Palestine advocators was to support the East Africa plan by presenting it as part of a "reversed" colonial plan. This plan meant establishing colonies as a temporary solution and only later founding a Jewish metropole in Palestine. Wilbush mentioned this plan in his memoir and it is possible that he considered this option while exploring Guas Ngishu. See Adam Rovner, *In the Shadow of Zion: Promised Lands before Israel* (New York: NYU Press, 2014), 57; Alroey, "Journey to New Palestine"; Nahum Wilbush, *Ha-Masa Le-Uganda* (Jerusalem: Ha-Sifriya Ha-Tzionit, 1963), 11.

24. Letter from Warburg to Weizmann, November 14, 1911, CZA, A12/71.

25. Wilbush, *Ha-Masa Le-Uganda*, 26; Alroey, "Journey to New Palestine," 36.

26. Ethan Katz, Lisa Moses Leff, and Maud Mandel, "Introduction: Engaging Colonial History and Jewish History, in *Colonialism and the Jews*, ed. Ethan Katz, Lisa Moses Leff, and Maud Mandel (Bloomington: Indiana University Press, 2017), 11.

27. Max Blanckenhorn, "Zum Klima von Palästina," *Zeitschrift des Deutschen Palästina-Vereins* 33, no. 2/3 (1910): 107–8; Yaron Perry and Efraim Lev, *Modern Medicine in the Holy Land: Pioneering British Medical Services in Late Ottoman Palestine* (London: Tauris Academic Studies, 2007), 68, 73. Scottish physicians were also the first to conduct meteorological investigation in India. Raj Kapil explains that while English universities in the nineteenth century (especially Oxford and Cambridge) focused on classical studies, Scottish universities increasingly emphasized science, law, and medicine and thus a large number of British colonial service positions were held by Scottish men or those trained at Scottish universities ("Colonial Encounters and the Forging of New Knowledge and National Identities: Great Britain and India, 1760–1850," *Osiris* 15 [2000]: 123–24).

28. Barbara W. Tuchman, *Bible and Sword: England and Palestine from the Bronze Age to Balfour* (London: Papermac, 1982), viii.

29. Eitan Bar-Yosef, *The Holy Land in English Culture 1799–1917: Palestine and the Question of Orientalism* (Oxford: Clarendon, 2005), 3, 7.

30. Jacob Norris, *Land of Progress: Palestine in the Age of Colonial Development, 1905–1948* (Oxford: Oxford University Press, 2013); Zvi Shilony, *Jewish National Fund and Settlement in Eretz-Israel 1903–1914* [Hebrew] (Jerusalem: Yad Ben-Zvi, 1990), 39. Besides a few nomadic Bedouin tribes that sojourned in the region seasonally, this region was scarcely settled and it was owned by the Ottoman state, which meant that it could be leased more easily than from private owners.

31. Clemens Wachter, *Die Professoren und Dozenten der Friedrich-Alexander-Universität Erlangen 1743–1960 (Teil 3)* (Erlangen: Universitätsbund Erlangen-Nürnberg, 2009), 19. According to other sources, Blanckenhorn's first surveys in Palestine were undertaken in 1894 at the invitation of Theodor Herzl; see Norris, *Land of Progress*, 60.

32. Max Blanckenhorn, "Studien über das Klima des Jordantals," *Zeitschrift des Deutschen Palästina-Vereins* 32, no. 1/2 (1909): 40.

33. Blanckenhorn, "Zum Klima von Palästina," 109.

34. Norton M. Wise, *The Values of Precision* (Princeton: Princeton University Press, 1995), 9.

35. Blanckenhorn, "Zum Klima von Palästina," 109.

36. Letter from Aaronsohn to the CEP, 07/02/1905, CZA, L1\53; and letter from Aaronsohn to Warburg, 05/08/1910, CZA L1/66.

37. Letter from Avraham Baruch (Rosenstein) to the management of the Ben Shemen farm, 26/12/1913, CZA, KKL3/859.

38. Dov Ashbel, *Avir Artzenu* (Tel Aviv: Omanut Ba'am, 1940), 15–17. Also see for example a letter from Raanan Weitz to Dov Ashbel about the financial aid of the JA in establishing a new meteorological station in Kibboutz Hanita, 29/05/1938, CAHUJI, 120/239.

39. Fredrik Meiton, *Electrical Palestine: Capital and Technology from Empire to Nation* (Oakland: University of California Press, 2019), 69–70; Shira Pinhas, "Road, Map: Partition in Palestine from the Local to the Transnational," *Journal of Levantine Studies* 10, no. 1 (2020): 112.

40. Pinhas, "Road, Map," 112.

41. Giuditta Parolini, "Weather, Climate, and Agriculture: Historical Contributions and Perspectives from Agricultural Meteorology," *WIREs Climate Change* 13, no. 3 (2022): 1–19.

42. Nebeker, *Calculating the Weather*, 83.

43. Ibid., 83–84.

44. Ibid.

45. Ashbel, *Avir Artzenu*, 13.

46. David E. Omissi, *Air Power and Colonial Control: The Royal Air Force 1919-1939* (Manchester: Manchester University Press, 1990), 44–45.

47. Yossi Malki, "Moderniyut, Leumiyut Ve-hevra: Reshit Ha-teufa Ha-ivrit Be-palestina- Eretz Israel" (PhD diss., Tel Aviv University, 2007), 17; Building the airport actually stood in contrast to the British usual passive administration in the country. While pursuing its obligations as required by the League of Nations, when possible, the British often preferred letting the Zionists invest in the country's development—especially in the fields of medical institutions, irrigation systems, and building plans—thus indirectly supporting and enhancing the segregation and imbalanced conditions between the Jewish and Arab sectors in the country. See M. Levy, *Merkaz Ha-Chizuyi Ve-Pe'ulotav* 16 (Beit Dagan: Ha-Sherut Ha-Meteorologi, 1975), 1.

48. For Feige's curriculum vitae and other personal documents, see AGSJI, G.F.0341-3/ 1.

49. AGSJI, G.F.0341-3/ 1; Peter Fritzsche, *A Nation of Fliers: German Aviation and the Popular Imagination* (Cambridge, MA: Harvard University Press, 1992).

50. Information on the participation of Feige and his gliding team in the *Makabyah* is in AGSJI, G.F.0341-1.

51. "Das Segelflugzeug als meteorologisches Forschungsmittel," March 1935, AGSJI G.F. 341-16/15.

52. "Das jüedische Flugwesen in Eretz Israel," *Ha-Masad* 4, no. 12 (1939): 184; Malki, "Moderniyut," 6–7, 15–16, 26, 30; testimonies on the beginnings of

aviation in Palestine from Michael Uri, March 30, 1958, HHA, 33.00012, and Yosef Keren, unknown date, HHA, 122.00029.

53. Conversation with M. Gumbley, Director of Civil Aviation at Lydda Airport, September 6, 1938, CAHUJI, 120/239.

54. Ashbel, *Avir Artzenu*, 14–16. According to Meiton, a similar Arab objection to the electric grid during the 1920s did not reflect so much a rejection of modern technology as a rejection of the lines of the electric grid that delineated the future Jewish state, and represented British and Zionist colonial symbols (*Electrical Palestine*, 151–52).

55. Aaron Sandler, "The Health Conditions of Palestine," in *Zionist Work in Palestine*, ed. Israel Cohen (New York: Judaean Publishing Company, 1912), 74–75.

56. Ashbel, *Avir Artzenu*, 70–71.

57. Zali Gurevitch and Gideon Aran, "On Israeli and Jewish Place: Israeli Anthropology," in *On Israeli and Jewish Place* [Hebrew] (Tel Aviv: Am Oved, 2007).

58. Caroline Ford, "Nature, Culture and Conservation in France and Her Colonies 1840–1940," *Past & Present* 183 (2004): 174–75.

59. Caroline Ford, *Natural Interests: The Contest over Environment in Modern France* (Cambridge, MA: Harvard University Press, 2016), 7.

60. Svetlana Boym, *The Future of Nostalgia* (New York: Basic Books, 2001), xiii.

61. Theodor Zlocisti, *Klimatologie und Pathologie Palästinas* (Tel Aviv: Omanut Co. Ltd, 1937), 18.

62. Charles William Wilson, "On Recent Surveys in Sinai and Palestine," *Proceedings of the Royal Geographical Society of London* 17, no. 5 (1873): 328. Emphasis added.

63. Theodor Herzl, *Altneuland = Old-New Land: Novel* (Haifa: Haifa Pub. Co, 1960 [1902]), 103. Emphasis added.

64. Sandler, "The Health Conditions of Palestine," 73–75. Emphasis added.

65. Report by Hershkowitz Elhanan, *Hatza'ot Ha-Veada Le-Gidul Atzei Pri Sub-Tropiim* (Suggestions of the Committee for Sub-tropic Fruit Cultivation), 06/07/1944, PLILMR, IV-293–75. Emphasis added.

66. Deborah R. Coen, "Climate and Circulation in Imperial Austria," *The Journal of Modern History* 82, no. 4 (December, 2010): 840.

67. Bruno Latour, "How Better to Register the Agency of Things," The Tanner Lecture on Human Values 34 (2016 [March 26, 2014]): 86.

68. John Briggs et al., "The Nature of Indigenous Environmental Knowledge Production: Evidence from Bedouin Communities in Southern Egypt," *Journal of International Development* 19 (2007): 242–43.

69. Salim Tamari, *Mountain against the Sea : Essays on Palestinian Society and Culture* (Berkeley: University of California Press, 2009), 109. See also

Khaled Furani and Dan Rabinowitz, "The Ethnographic Arriving of Palestine," *Annual Review of Anthropology* 40 (2011): 475–91.

70. Tawfik Canaan, "Der Kalender des palästinensischen Fellachen," *Zeitschrift des Deutschen Palästina-Vereins* 36, no. 4 (1913): 276.

71. Tawfik Canaan, "Folklore of the Seasons in Palestine," *Journal of the Palestine Oriental Society* 3 (1923): 29.

72. Canaan, "Der Kalender des palästinensischen Fellachen," 287.

73. Ibid., 268.

74. Canaan, "Folklore of the Seasons in Palestine," 29.

75. Canaan, "Der Kalender des palästinensischen Fellachen," 296; Stephan Hanna Stephan, "The Division of the Year in Palestine," *The Journal of the Palestine Oriental Society* II (1922): 162.

76. Canaan, "Der Kalender des palästinensischen Fellachen," 270.

77. Stephan, "The Division of the Year in Palestine," 163.

78. The monthly and yearly statistics of the Fruit Growers' Union also make clear that until the 1930s Palestinians held much larger quantities of agricultural land and accordingly produced greater yields (usually more than double) than Jewish farmers. See PLILMR, IV293–81, IV-293–75. For further reading on this subject see also Amos Nadan, *The Palestinian Peasant Economy under the Mandate : A Story of Colonial Bungling* (Cambridge, MA: Harvard University Press, 2006).

79. Canaan, "Der Kalender des palästinensischen Fellachen," 270.

80. Fritz Kahn, *Naturgeschichte Palästinas*, c. 1930s, LBICJH, AR 7144/MF 746, II,4, 2:42. Emphasis added.

CHAPTER 2. CLIMATE AND THE JEWISH EUROPEAN BODY

1. For further reading on this subject in the early modern period, see Dror Wahrman, *The Making of the Modern Self : Identity and Culture in Eighteenth-Century England* (New Haven: Yale University Press, 2004), 83–126; Roxann Wheeler, *The Complexion of Race : Categories of Difference in Eighteenth-Century British Culture* (Philadelphia: University of Pennsylvania Press, 2000); Richard Grove, *Green Imperialism: Colonial Expansion, Tropical Island Edens and the Origins of Environmentalism, 1600–1860* (Cambridge, UK: Cambridge University Press, 1995).

2. Hippocrates's treatise *De aere, aquis, locis* (Air, Waters, and Places), from the fourth century BCE, is considered to be the first essay discussing the relation between climate, health, and ethnography.

3. Mark Harrison, "'The Tender Frame of Man': Disease, Climate, and Racial Difference in India and the West Indies, 1760–1860," *Bulletin of the History of*

Medicine 70 (1996): 68–93; and Mark Harrison, *Climates & Constitutions: Health, Race, Environment and British Imperialism in India, 1600–1850* (Oxford: Oxford University Press, 1999), 106.

4. John M. Efron, *Defenders of the Race: Jewish Doctors and Race Science in Fin-De-Siecle Europe* (New Haven: Yale University Press, 1994), 9.

5. Mitchell Bryan Hart, *Social Science and the Politics of Modern Jewish Identity* (Stanford, CA: Stanford University Press, 2000), 21. Other studies that focus on Jewish-Zionist racial thought during the twentieth century include Efron, *Defenders of the Race*; Raphael Falk, *Zionism and the Biology of the Jews* (New York: Springer, 2017); and Dafna Hirsch, "Zionist Eugenics, Mixed Marriage, and the Creation of a 'New Jewish Type'," *The Journal of the Royal Anthropological Institute* 15, no. 3 (2009).

6. Elias Auerbach, *Palästina als Judenland* (Berlin, Leipzig: Aktionskomitee der zionistischen Organisation, in Kommission beim juedischen Verlag, 1912); Arthur Ruppin, "The Selection of the Fittest," in *Three Decades of Palestine: Speeches and Papers on the Upbuilding of the Jewish National Home* (Jerusalem: Schocken, 1936); and Todd Endelman, "Anglo-Jewish Scientists and the Science of Race," *Jewish Social Studies* 11, no. 1 (Autumn 2004): 72. Dafna Hirsch has claimed that this approach was especially common among German Jews, who advocated for their own reform, freedom of choice, and progress; see *"We Are Here to Bring the West": Hygiene Education and Culture Building in the Jewish Society of Palestine During the British Mandate Period* [Hebrew] (Sde Boker: The Ben-Gurion Research Institute for the Study of Israel and Zionism, 2014), 88–89.

7. Henry Thomas Buckle, *History of Civilisation in England* (London: John W. Parker & Son, 1857).

8. Woodruff D. Smith, "Friedrich Ratzel and the Origins of Lebensraum," *German Studies Review* 3, no. 1 (1980): 51–68; Friedrich Ratzel, *Politische Geographie* (München und Leipzig: R. Oldenbourg 1897).

9. Vladimir Zeev Jabotinsky, "Ziyonut Ve-Eretz Israel" *Ktavim Tziyoniyim Rishonim* (Jerusalem: A. Jabotinsky, 1905), 123–24.

10. Ibid., 111. To support his argument, Jabotinsky also referred to the work of Buckle.

11. Quoted in Dana von Suffrin, *Pflanzen für Palästina: Otto Warburg und die Naturwissenschaften im Jischuw* (Tübingen: Mohr Siebeck, 2019), 86.

12. Otto Warburg, "Die nichtjüdische Kolonisation Palästinas," *Altneuland* 2 (February 1904): 43.

13. Von Suffrin, *Pflanzen für Palästina*, 87.

14. David Arnold, *Warm Climates and Western Medicine: The Emergence of Tropical Medicine, 1500–1900* (Amsterdam: Rodopi, 1996), 9–10.

15. Eric T. Jennings, *Curing the Colonizers: Hydrotherapy, Climatology, and French Colonial Spas* (Durham: Duke University Press, 2006), 21, 28–29.

16. Efron, *Defenders of the Race*, 56.

17. Jennings, *Curing the Colonizers*, 29.

18. Amnon Raz-Krakotzkin, "En Elohim, Aval Hu Hevtiach Lanu Et Ha-Aretz," *Mita'am* 3 (2005): 72.

19. Quoted in von Suffrin, *Pflanzen für Palästina*, 86.

20. Raphael Falk, "Zionism and the Biology of the Jews," *Science in Context* 11, no. 3-4 (1998): 595.

21. Efrat Gilad, "Meat in the Heat: A History of Tel Aviv under the British Mandate for Palestine (1920s-1940s)," PhD diss., University of Geneva, 2021, 90–91.

22. Jakob Tahon, *Warburg's Book* [Hebrew] (Tel Aviv: Masada, 1948), 134.

23. Theodor Zlocisti, *Klimatologie und Pathologie Palästinas* (Tel Aviv: Omanut Co. Ltd, 1937), 36–38.

24. Falk, "Zionism and the Biology of the Jews," 595.

25. On Barak, *Powering Empire : How Coal Made the Middle East and Sparked Global Carbonization* (Oakland: University of California Press, 2020), 16–17.

26. Eliezer Livneh, *Aaron Aaronsohn: Ha-Ish Ve-Zmano* (Jerusalem: Bialik, 1969), 61.

27. Sander L. Gilman, *The Jew's Body* (London: Routledge, 1991), 171–73. This claim is also made by Efron (*Defenders of the Race*, 50–51).

28. Robert Knox, *The Races of Men : A Fragment* (London: Henry Renshaw, 1850), 133–34.

29. Gilman, *The Jew's Body*, 174.

30. Cited in Daniel Boyarin, *Unheroic Conduct: The Rise of Heterosexuality and the Invention of the Jewish Man* (Berkeley: University of California Press, 1997), 263.

31. Theodor Herzl, *The Complete Diaries of Theodor Herzl*, ed. Raphael Patai (New York: Herzl Press, 1961), 1:276, entry for November 21, 1896.

32. Axel Stähler, *Zionism, the German Empire, and Africa: Jewish Metamorphoses and the Colors of Difference* (Oldenbourg: De Gruyter, 2019), 42.

33. Homi K. Bhabha, "The White Stuff (Political Aspect of Whiteness)," *Artforum* 36, no. 9 (1998): 21–24.

34. Boyarin, *Unheroic Conduct*.

35. Malaria was only one out of many infectious diseases in Palestine during this time. Other contemporary local diseases included cholera, typhus, influenza, diphtheria, dysentery, measles, rubella, mumps, and trachoma. However, malaria killed the largest number of people and was perceived as the biggest problem.

36. Sandra M. Sufian, *Healing the Land and the Nation: Malaria and the Zionist Project in Palestine, 1920-1947* (Chicago: University of Chicago Press, 2007).

37. Alan Mikhail, *Nature and Empire in Ottoman Egypt : An Environmental History* (Cambridge: Cambridge University Press, 2011), 201–41; and Khaled Fahmy, "An Olfactory Tale of Two Cities: Cairo in the Nineteenth Century," in *Historians in Cairo: Essays in Honor of George Scanlon*, ed. Jill Edwards (Cairo: American University in Cairo Press, 2002), 167.

38. John Farley, *Bilharzia: A History of Imperial Tropical Medicine* (Cambridge, UK: Cambridge University Press, 1991); Dane Kennedy, "The Perils of the Midday Sun: Climatic Anxieties in the Colonial Tropics," in *Imperialism and the Natural World*, ed. John M. MacKenzie (Manchester: Manchester University Press, 1990), 118–40; Anna Crozier, "Sensationalising Africa: British Medical Impressions of Sub-Sahara Africa, 1890–1939," *Journal for Imperial and Commonwealth History* 35, no. 3 (2007): 393–415; N. David Livingstone, "Race, Space and Moral Climatology: Notes Towards a Genealogy," *Journal of Historical Geography* 28, no. 2 (2002): 159–80; and Ryan Johnson, "Networks of Imperial Tropical Medicine: Ideas and Practices of Health and Hygiene in the British Empire, 1895–1914" (PhD diss., University of Oxford, 2009).

39. Jennings, *Curing the Colonizers*, 33.

40. B.S., "Gam Elle Goalim," *He-Ahdut*, July 10, 1914, 6.

41. Ibid. Another similar story mentions that the proximity of a certain Jewish settlement to swamps made many of its members sick while having "green faces"; see *Zikhronot Me-Eretz Israel*, ed. Yehoshua Bivar (Jerusalem: Ha-Machlaka Le-Chinukh U-Le-Tarbut Ba-Gola Shel Ha-Histadrut Ha-Tziyonit, 1985), 7–8, 14.

42. *Zikhronot Me-Eretz Israel*, 7–8.

43. Aaron Sandler, "The Health Conditions of Palestine," in *Zionist Work in Palestine*, ed. Israel Cohen (New York: Judaean Publishing Company, 1912), 73.

44. Reports on the Gaza district villages, 1920–1921, CZA, L18/6290.

45. Diana K. Davis, *The Arid Lands: History, Power, Knowledge* (Cambridge, MA: MIT Press, 2016), 33.

46. Johnson, "Networks of Imperial Tropical Medicine," 2, 5. This idea is supported by scholarship including Michael Worboys, "The Colonial World as Mission and Mandate: Leprosy and Empire, 1900–1940," *Osiris* 15, no. 1 (2000): 207–18; David Arnold, *Imperial Medicine and Indigenous Societies* (Manchester, UK: Manchester University Press, 1988), 2; Farley, *Bilharzia*, 4; and Megan Vaughan, *Curing Their Ills : Colonial Power and African Illness* (Cambridge, UK: Polity, 1991).

47. Johnson, "Networks of Imperial Tropical Medicine," 5.

48. Anne Marie Moulin, "Tropical without the Tropics: The Turning-Point of Pastorian Medicine in North Africa," in *Warm Climates and Western Medicine: The Emergence of Tropical Medicine, 1500–1900*, ed. David Arnold (Amsterdam: Rodopi, 1996), 160–80.

49. Anti-Malaria Measures—Malaria Pamphlet, ISA 6542/28-מ. 1924.

50. D. Beckmann, "Eikh Le-Hishamer Mi-Malaria," in *Briut Ha-Oleh: Madrikh La-Oleh Ha-Chadash Be-Inyanei Higyena U-Vriut* (Jerusalem: Kupat Cholim Amamit Ve-Machleket Ha-Aliya Shel Ha-Sokhnut Ha-Yehudit, 1939), 13–14; and Jacob Seide, *Briut Ha-Am: Ha-Teva, Ha-Aklim, Ve-Ha-Briut Be-Israel* (Jerusalem: Reuven Mas, 1954), 37.

51. Israel Jacob Kligler, *The Epidemiology and Control of Malaria in Palestine* (Chicago: University of Chicago Press, 1930), I.

52. Ibid., 15. Tawfik Canaan similarly argued that the Palestinian diet, which was usually vegetarian and based on legumes, was good for a strong immune system, in "Aberglaube und Volksmedizin im Lande der Bibel," *Abhandlungen des Hamburgischen Kolonialinstituts* 20, no. 3 (1914): 1–2, 5–6, 10. While Jewish health institutions were occupied with settlers' diet from the 1930s, British authorities only gave attention to local Arab diet in the 1940s. A special unit for nutrition was established in the Department of Health in 1942. See Sandra M. Sufian, "Arab Health Care During the British Mandate, 1920–1947," in *Separate and Cooperate, Cooperate and Separate: The Disengagement of the Palestine Health Care System from Israel and Its Emergence as an Independent System*, ed. Tamara Barnea and Rafiq Husseini (London: Praeger, 2002), 12.

53. Kligler, *The Epidemiology and Control of Malaria in Palestine*, 12–13.

54. Diana K. Davis and Edmund Burke III, eds., *Environmental Imaginaries of the Middle East and North Africa* (Athens: Ohio University Press; London, 2011), 12.14–16.

55. Sufian, *Healing the Land and the Nation*, 146. It was in this spirit that Ottoman efforts to drain swamps in the country during the nineteenth century were ignored, making the British (and, in other cases, the Zionists) seem like the precursors of progress.

56. As Timothy Mitchell reminds us, in 1915, one year after the completion of the Panama Canal, the newly founded Rockefeller Foundation took over the mosquito campaign from the US army and launched a worldwide program to study and control mosquito-borne diseases. As Mitchell puts it, the global movements of the mosquito gave shape to a transnational corporate philanthropy. See *Rule of Experts: Egypt, Techno-Politics, Modernity* (Berkeley: University of California Press, 2002), 30.

57. Herbert Samuel, *Report of the High Commissioner on the Administration of Palestine 1920–1925* (London: His Majesty's Stationery Office, 1925), 10–11.

58. Sufian, *Healing the Land and the Nation*, 208; Zalman Greenberg and Anton Alexander, "Israel Jacob Kligler: The Story of a 'Little Big Man', a Giant in the Field of Public Health in Palestine," *Korot* 21 (2011–2012): 176–78.

59. Sufian, *Healing the Land and the Nation*, 138–41.

60. Ibid; and Greenberg and Alexander, "Israel Jacob Kligler," 196.

61. Warwick Anderson, *The Cultivation of Whiteness: Science, Health and Racial Destiny in Australia* (Carlton, Australia: Melbourne University Press, 2005), 4.

62. Rabi Zeira in the Babylonian Talmud, *Bavra Batra* 158: 72.

63. Hilel Yaffe, *Shmirat Ha-Briut* (Jaffa: A. Atin Print, 1913), 2.

64. Tamar Novick, *Milk and Honey: Technologies of Plenty in the Making of a Holy Land* (Cambridge, MA: MIT Press, 2023).

65. Shmuel Yosef Agnon, *Only Yesterday* (Princeton: Princeton University Press, [1945] 2000), 25.

66. Jennings, *Curing the Colonizers*, 39, 49–50.

67. Mirjam Triendl-Zadoff, *Next Year in Marienbad : The Lost Worlds of Jewish Spa Culture* (Philadelphia: University of Pennsylvania Press, 2012), 32–34.

68. Ibid., 33–34, 37–38.

69. Zlocisti, *Klimatologie und Pathologie Palästinas*, 18.

70. Yael Levy and Nissim Levy, *The Physicians of the Holy Land, 1788-1948* [Hebrew](Zikhron Ya'akov: Itay Bakhor, 2017), 62.

71. Hirsch, *"We Are Here to Bring the West."*

72. Zlocisti, *Klimatologie und Pathologie Palästinas*, 45.

73. Ibid., 44.

74. Ibid.

75. Dafna Hirsch, "'We Are Here to Bring the West, Not Only to Ourselves': Zionist Occidentalism and the Discourse of Hygiene in Mandate Palestine," *International Journal of Middle East Studies* 41, no. 4 (2009).

76. Sufian, *Healing the Land and the Nation*, 11.

77. Sufian, "Arab Health Care During the British Mandate, 1920–1947," 14, 15; and Sufian, *Healing the Land and the Nation*, 12.

78. Hirsch, "'We Are Here to Bring the West, Not Only to Ourselves,'" 580.

79. Ibid., 583–84; and Sufian, *Healing the Land and the Nation*, 25.

80. Quoted in Ori Yehudai, "Displaced in the National Home: Jewish Repatriation from Palestine to Europe, 1945–48," *Jewish Social Studies* 20, no. 2 (Winter 2014): 95.

81. Quoted in Ori Yehudai, *Leaving Zion: Jewish Emigration from Palestine and Israel after World War 2* (Cambridge, UK: Cambridge University Press, 2020), 76.

82. Sandler, "The Health Conditions of Palestine," 73.

83. *Beriut Ha-Oleh: Madrich La-Oleh He-Hadash Be-Inyanei Higyena U-Vriut* (Jerusalem: Kupat Holim Amamit Ve-mahleket Ha-aliya Shel Hasokhnut Ha-yahudit, 1939), 19.

84. Ibid., 4.

85. Seide, *Briut Ha-Am*, 7.

86. Ibid.; R. Kazenelson, *Briut Ha-Ole: Madrikh La-Ole Ha-Chadash Be-Inyanei Higyena U-Vriut* (Jerusalem: Kupat Cholim Amamit Ve Machleket Ha-Aliya Shel Ha-Sokhnut Ha-Yehudit, 1939); and Abraham Jehuda Ha-levi, *Briut Ha-Tzibur* (Tel Aviv: Achiever, 1935).

87. Seide, *Briut Ha-Am*, 28–29, 35.

88. Zlocisti, *Klimatologie und Pathologie Palästinas*, 74.

89. Dr. A. Binyamini, "Ha-Bechinot U-Veriut Yeladeinu," *Ha-Aretz*, July 24, 1930, 3.

90. Walter Strauss, "Ha-Aklim Ha-Cham Ve-Ha-Tilboshet," in *Ha-Adam Ve-Ha-Aklim Be-Eretz-Israel*, ed. S. Rosenbaum (Tel Aviv: Olympia, n.d.), 57.

91. M. Shechter, "Aseret Ha-Dibrot Le-Yemei Ha-Hamsin," *Davar*, May 6, 1935, 8.

92. Luise Hirsch, *From the Shtetl to the Lecture Hall: Jewish Women and Cultural Exchange* (Lanham, MD: University Press of America, 2013), 251.

93. Erna Meyer, *How to Cook in Palestine?* (Tel Aviv: Palestine Federation of WIZO, 1937), 7.

94. Ibid., 8.

95. Seide, *Briut Ha-Am*, 116–19. For further reading on climate and nutrition in Palestine during this period see Gilad, "Meat in the Heat."

96. Harmke Kamminga and Andrew Cunningham, *The Science and Culture of Nutrition, 1840–1940* (Amsterdam: Rodopi, 1995), 2.

97. Yael Raviv, *Falafel Nation : Cuisine and the Making of National Identity in Israel* (Lincoln: University of Nebraska Press, 2015), 96.

98. Exodus 16:8, cited in M. Shechter, "Al Ha-Ma'akhlim Ve Ha-Mashkaot," *Davar*, May 5, 1936, 9.

99. Gilad, "Meat in the Heat."

100. Quoted in Raviv, *Falafel Nation*, 96.

101. Quoted in Ofra Tene, "The New Immigrant Must Not Only Learn," in *Jews and Their Foodways*, ed. Anat Helman (Oxford: Oxford University Press, 2016), 51.

102. Ibid.

103. Zlocisti, *Klimatologie und Pathologie Palästinas*, 42; and Rudolf Feige, "He'arot Al Ha-Klimatologia Ha-Shimushit Be-Eretz Israel," in *Ha-Adam Veha-Aklim Be-Eretz Israel*, ed. S. Rosenbaum (Tel Aviv: Olympia, n.d.), 37.

104. Conevery Bolton Valencius, *The Health of the Country: How American Settlers Understood Themselves and Their Land* (New York: Basic Books, 2002), 27–28; and Johnson, "Networks of Imperial Tropical Medicine," 45–50, 78, 80.

105. Seide, *Briut Ha-Am*, 21.

106. Zlocisti, *Klimatologie und Pathologie Palästinas*, 166.

CHAPTER 3. WARM PALESTINIAN CLIMATE—COOL JEWISH SPACES

1. Many parts of this chapter are based on my MA thesis: Netta Cohen, "Environment, Climate and the East in the Eyes of Jewish Architects in Palestine 1909–1948" [Hebrew], Tel Aviv University, 2013.

2. Mike Hulme, "The Conquering of Climate: Discourses of Fear and Their Dissolution," *The Geographical Journal* 174, no. 1 (2008): 10.

3. See, for instance, Rebecca J. H. Woods, "Nature and the Refrigerating Machine: The Politics and Production of Cold in the Nineteenth Century," in *Cryopolitics : Frozen Life in a Melting World,* ed. Joanna Radin and Emma Kowal (Cambridge, MA: MIT Press, 2017); Jonathan Rees, *Refrigeration Nation: A History of Ice, Appliances, and Enterprise in America* (Baltimore: John Hopkins University Press, 2013); and On Barak, *Powering Empire: How Coal Made the Middle East and Sparked Global Carbonization* (Oakland: University of California Press, 2020).

4. Woods adds that refrigerating technologies also liberated colonists from the cumbersome implications of seasons and geographical limits. As one contemporary individual indicated, the refrigerating industry "promised to seemingly unite the hemispheres—to weave the northern and the southern domains of the British Empire into one frictionless whole, transcended climatological, cultural or even seasonal differences" ("Nature and the Refrigerating Machine," 95, 107).

5. Barak, *Powering Empire,* 58.

6. Theodor Herzl, *Altneuland = Old-New Land: A Novel* (Haifa: Haifa Pub. Co, 1960 [1902]), 126.

7. Gail Cooper, *Air-Conditioning America: Engineers and the Controlled Environment, 1900–1960* (Baltimore: Johns Hopkins University Press, 1998), 1.

8. Barak, *Powering Empire,* 16–17.

9. In Cooper, *Air-Conditioning America,* 8.

10. Ibid.

11. Fritz Kahn, *Naturgeschichte Palästinas,* c. 1930s, LBICJH, AR 7144/MF 746, II,4, 2:42.

12. In North America air conditioning made its first substantial appearance in 1899 in central stations in several cities including New York, Boston, Baltimore, Los Angeles, St. Louis, Denver, and Kansas City. Later, pipes from these central stations were extended into nearby restaurants and other businesses. In 1902, the same year in which Herzl published his utopian novel, the New York Stock Exchange installed an air conditioning system in one of its new buildings, and by 1904 the American Society of Heating and Ventilating Engineers was founded (Cooper, *Air-Conditioning America,* 8–10).

13. "Mizug Avir Yutkan Be-Ein Gedi," *Al Ha-Mishmar,* July 22, 1956, 4; "Israel Meyatzet Memazgei Avir," *She'arim,* July 13, 1956, 8; and "Memazgei Avir Me-Israel Le-Rachavei Ha-Olam," *Ma'ariv,* July 12, 1956.

14. Woods, "Nature and the Refrigerating Machine," 89, 99; and Rees, *Refrigeration Nation,* 1, 4.

15. "Beit Ha-Charoshel Le-Kerach Tel Aviv," *Ha-Aretz,* May 3, 1928.

16. Barak, *Powering Empire,* 67.

17. "Beit Ha-Charoshel Le-Kerach Tel Aviv."

18. Efrat Gilad, "Meat in the Heat: A History of Tel Aviv under the British Mandate for Palestine (1920s-1940s)" (PhD diss., University of Geneva, 2021), 89.

19. "Elef Tzarkhanei Kerach Be-Tel Aviv," *Ha-Mashkif,* June 10, 1946, 40.

20. Sherene Seikaly, *Men of Capital : Scarcity and Economy in Mandate Palestine* (Stanford, CA: Stanford University Press, 2016), 37.

21. "Haifa Shuv Sha'aruruyat Kerach," *Ha-Tzofe,* August 29, 1945.

22. Untitled, *Ha-Mashkif,* July 15, 1945, 4.

23. Issum Perot Be-Onat Ha-Kaitz, March 1945, PLILMR, IV-293-75.

24. Fruit Growers' Union handbook from September 1942, PLILMR, IV-293-74.

25. Etan Bloom, *Arthur Ruppin and the Production of Pre-Israeli Culture* (Leiden: Brill; Biggleswade, 2011), 156.

26. Sir Howard Ebenezer, *Garden Cities of To-Morrow* (London: Swan Sonnenschein & Co., Ltd, 1902); and Josef Stübben, *Handbuch der Architektur: der Stadtbau* (Darmstadt: Arnold Bergstrasser, 1890).

27. Zvi Efrat, *The Israeli Project: Building and Architecture 1948-1973* [Hebrew] (Tel Aviv: Tel Aviv Museum of Art, 2004).

28. Tovi Fenster, *Whose City Is It? Planning, Knowledge and Everyday Life* [Hebrew] (Tel Aviv: Ha-Kibbutz Ha-Meuchad, 2012), 31; Haim Yacobi, "Words and Place: The Case of the City Lod," in *Architectural Culture: Place, Representation, Body* [Hebrew], ed. Tali Hatuka and Rachel Kallus (Tel Aviv: Resling, 2005), 85; Mark Levine, *Overthrowing Geography: Jaffa, Tel Aviv and the Struggle for Palestine 1880-1948* (Oakland: University of California Press, 2005), 15; and Robert Home, *On Planting and Planning: The Making of British Colonial Cities* (London: Taylor & Francis, 1997), 125-26.

29. Fenster, *Whose City Is It?*; and Yacobi, "Words and Place."

30. Letter from Albert Klein to his family, November 14, 1935, LBIJER 39.

31. Miki Zaidman and Ruth Kark, "The Beginnings of Tel Aviv: The Achuzat Bayit Neighborhood as a 'Garden City'?" [Hebrew], *Zmanim: A Historical Quarterly* 106 (Spring 2009): 13.

32. Quoted in Dafna Hirsch, *'We Are Here to Bring the West': Hygiene Education and Culture Building in the Jewish Society of Palestine During the British Mandate Period* [Hebrew] (Sde Boker: The Ben-Gurion Research Institute for the Study of Israel and Zionism, 2014), 12.

33. Levine, *Overthrowing Geography.*

34. Sonja Duempelmann, *Seeing Trees : A History of Street Trees in New York City and Berlin* (New Haven: Yale University Press 2019), 25, 120.

35. Noah Hysler-Rubin, "A Cultural Acropolis in Tel Aviv: Patrick Geddes' Vision for the First Hebrew City" [Hebrew], *Zmanim: A Historical Quarterly* 106 (Spring, 2009): 36; and Ofra Tene, "The White Houses Were Filled: Everyday Life in Homes in Tel Aviv 1924–1948" [Hebrew] (PhD diss., Tel Aviv University, 2010), 32–33.

36. Noah Hysler-Rubin, *Patrick Geddes and Town Planning: A Critical View* (London: Routledge, 2011), 36–38.

37. Patrick Geddes, *Town-Planning Report—Jaffa and Tel Aviv* (1925), 19.

38. Nitza Metzger-Szmuk, *Dwelling on the Dunes: Tel Aviv, Modern Movement and Bauhaus Ideals* (Paris: L'Éclat, 2004), 17–23.

39. Dov Karmi, "Ha-Orientatzia Shel Ha-Dira Be-Tel Aviv," *Building in the Near East* 9–10 (November 1936): 5.

40. "Le-Takanat Ha-Batim Ha-Nivnim Be-Tel-Aviv," in *Tel Aviv—a Silver Jubilee: A Documentary Anthology* [Hebrew], ed. Maoz Azaryahu, Arnon Golan, and Aminadav Dikman (Jerusalem: Carmel Publication, [October 15, 1935] 2009), 180.

41. Dov Karmi, "Mibe'ayot Tikun Ha-Dirot Be-Tel Aviv," *Journal of the Association of Engineers and Architects* 7 (1945): 9.

42. Yuval Yaski and Galia Bar Or, *Kibbutz: Architecture without Precedents* [Hebrew] (Ein Harod: Museum of Art, 2010), 270, 341–42; and Elissa Rosenberg, "An All Day Garden: The Kibbutz as a Modernist Landscape," *Journal of Landscape Architecture* 7, no. 2 (2012): 33.

43. Yaski and Bar Or, *Kibbutz*, 322, 273, 78.

44. Muky Tsur and Yuval Danieli, *Mestechkin Builds Israel: Architecture in the Kibbutz* [Hebrew] (Tel Aviv: Ha-kibbutz Ha-Meuhad, 2008).

45. Freddy Kahana, *Neither Town nor Village: The Architecture of the Kibbutz 1910–1990* [Hebrew] (Rishon Le-tzion: Ha-moatza Le-shimur Atarim, 2011), 65.

46. Tsur and Danieli, *Mestechkin Builds Israel*, 136.

47. Efrat, *The Israeli Project*, 59.

48. Alexander Berwald, "Bauliche Probleme in Palästina," *Die Welt* 41 (October, 1910).

49. Ibid., 1049.

50. Tawfik Canaan, *The Palestinian Arab House, Its Architecture and Folklore* (Jerusalem: Syrian Orphanage Press, 1933), III.

51. Salim Tamari, *Mountain against the Sea: Essays on Palestinian Society and Culture* (Berkeley: University of California Press, 2009), 109. See also Khaled Furani and Dan Rabinowitz, "The Ethnographic Arriving of Palestine," *Annual Review of Anthropology* 40 (2011): 475–91.

52. Josef Awin, "Al Omanut Ha-Bniya Be-Eretz Israel," *Binyan Ve-Charoshet* 6, no. 3–4 (1927): 3.

53. In Gil Gordon, "Gagot Mito'fefim Baruach: Knisatam Shel Re'afim Ve-Ta'asiyat Ha-Charsit Le-Eretz Israel," *Zmanim: A Historical Quarterly* 96 (Spring 2006): 61.

54. Berwald, "Bauliche Probleme in Palästina," 1048.

55. Ibid.

56. Ibid., 1049.

57. Or Aleksandrowicz, "Mishelet Churban: Chayav U-Moto Shel Binyan Ha-Gymnasia Hertseliya," in *Yemei Ha-Gymnasia*, ed. Guy Raz (Tel Aviv: Eretz Israel Museum, 2013), 32.

58. Çelik Zeynep, *Urban Forms and Colonial Confrontations: Algiers under French Rule* (Berkeley: University of California Press, 1997), 8.

59. "Le-Takanat Ha-Batim Ha-Nivnim Be-Tel-Aviv," 176.

60. J. H. Chang and A. D. King, "Towards a Genealogy of Tropical Architecture: Historical Fragments of Power-Knowledge, Built Environment and Climate in the British Colonial Territories," *Singapore Journal of Tropical Geography* 32, no. 3 (2011): 287.

61. Arab architects who built houses in Palestine according to the modernist style were similarly ignored. See Sharon Rotbard, *White City, Black City: Architecture and War in Tel Aviv and Jaffa* (London: Pluto Press, 2015), 60.

62. Tropical architecture is usually associated with an architectural style prevalent in the second half of the twentieth century. Chang and King, nevertheless, identify its earlier roots and principles in architectural ideas and methods dating from the mid-eighteenth century ("Towards a Genealogy of Tropical Architecture," 283).

63. Ibid.

64. Ibid., 287–88.

65. Alona Nitzan-Shiftan, "Disputes in Zionist Architecture Erich Mendelssohn and the Tel Aviv Hug," in *Architectural Culture: Place, Representation, Body* [Hebrew], ed. Rachel Kallus and Tali Hatuka (Tel Aviv: Resling, 2005), 206; and Anat Helman, *Urban Culture in 1920s and 1930s Tel Aviv* [Hebrew] (Haifa: Haifa University Press, 2007), 25.

66. Mark Crinson, *Modern Architecture and the End of Empire* (Aldershot: Ashgate, 2003), 2.

67. Rotbard, *White City, Black City*, 32.

68. Shlomo Ginzburg, "Ha-Orientatzya Shel Ha-Dira Be-Haifa," *Building in the Near East* 9–10 (November 1936): 6.

69. Avraham Haim Elhanani, "Bniya Ve-Tikhnun Lefi Aklim Ha-Aretz," *Davar*, July 20, 1949, 8.

70. Otto Schiller, "Ha-Avodot Ha-Mukdamot Letikhun Irony," *Building* 2 (1935): 1.

71. Ginzburg, "Ha-Orientatzya Shel Ha-Dira Be-Haifa," 7.

72. Karmi, "Ha-Orientatzia Shel Ha-Dira Be-Tel Aviv," 5.

73. Tene, "The White Houses Were Filled," 86.

74. Karmi, "Mibe'ayot Tikun Ha-Dirot Be-Tel Aviv," 11. Emphasis added.

75. In Or Aleksandrowicz, "Appearance and Performance: Israeli Building Climatology and Its Effect on Local Architectural Practice (1940–1977)," *Architectural Science Review* 60, no. 5 (2017): 372–73.

76. Ibid.

77. Dr. Groschka, "Ha-Dira Ba-Aklim Ha-Ham," in *Ha-Adam Ve-Ha-Aklim Be-Eretz Israel*, ed. S. Rosenbaum (Tel Aviv: Olympia, n.d.), 67.

78. Tene, "The White Houses Were Filled," 96.

79. Ibid.

80. David Kroyanker, *Jerusalem Architecture: Periods and Styles of the Period of the British Mandate 1918–1948* [Hebrew] (Tel Aviv: Keter, 1989), 44.

81. "Le-Takanat Ha-Batim Ha-Nivnim Be-Tel-Aviv," 178.

82. Julius Posener, "Batim Bnei Dira Achat Be-Eretz Israel," *Building* 1 (1934): 1.

83. Ibid.

84. Nahum Gutman, "Mashehu Al Tzurata Ha-Chitztonit Shel Ha-Ir," in *Tel Aviv—a Silver Jubilee: A Documentary Anthology* [Hebrew], ed. Maoz Azaryahu, Arnon Golan, and Aminadav Dikman (Jerusalem: Carmel Publishing, [1933] 2009), 129. Fritz Kahn addressed this same issue when he described the necessity of wearing sunglasses in Palestine, which according to him were indispensable for maintaining sight and comparable to the necessity of an umbrella in the English weather. See "Die orientalischen Augenkrankheiten" in *Naturgeschichte Palästinas,* c. 1930s, LBICJH, AR 7144/MF 746, II,4, 2:42.

85. "Le-Takanat Ha-Batim Ha-Nivnim Be-Tel-Aviv," 179.

86. Or Aleksandrowicz, *Daring the Shutter: The Tel Aviv Idiom of Solar Protections* (Tel Aviv: Public School, 2015), 7, 11, 22.

87. Or Aleksandrowicz, "Kurkar, Melet, Aravim, Jehudim: Eikh Bonim Ir Ivrit," *Teorya U-Vikoret (Theory and Criticism)* 36 (2010): 62.

88. David Kroyanker, *Jerusalem Architecture: Periods and Styles of Jewish Neighborhoods and Public Buildings Outside the City Walls [Hebrew]* (Tel Aviv: Keter, 1988), 27.

89. T. Tinovitch, "Hitpathut Ha-Taasiya Shel Homrei Binyan Ba-Aretz " *Journal of the Association of Engineers and Architects* 7 (1945): 3.

90. Ibid., 5.

91. Aleksandrowicz, "Kurkar, Melet, Aravim, Jehudim," 68.

92. Gershon Shafir, *Land, Labor and the Origins of the Israeli-Palestinian Conflict, 1882–1914* (Berkeley: University of California Press, 1996), 8.

93. Nimrod Ben Zeev, "Foundations of Inequality: Construction, Political Economy, Race and the Body in Palestine/Israel, 1918–1973" (PhD diss., The University of Pennsylvania, 2020), 3. Ben Zeev argues that despite the obvious

differences between the two communities in Palestine, the aim to "build a nation" in the material sense was also shared by Palestinian investors and builders who acknowledged the growing popularity of cement in their work and by the late 1920s began to see these materials as both an economic opportunity as well as an important factor in Palestinian national emancipation (51).

94. Hadas Shadar, *Sunstroke: Brutalist Construction in Be'er Sheva Re-Examination of National Architecture* [Hebrew] (Jerusalem: Yad Ben Zvi), 11.

95. Natan Alterman, "Shir Boker" *[Morning Song] Pizmonim Ve-Shirei Zemer* (Tel Aviv: Ha-Kibbutz Ha-Meuchad, 1979 [1934]), 2:302–3. Translated by Nili Schneidman. For further reading about the introduction of cement to Palestine and how it impacted the intricated relationship between Arab and Jews in Palestine see Ben Zeev, "Foundations of Inequality," 38–71.

96. Shadar, Sunstroke, 7.

97. Quoted in Efrat, *The Israeli Project*, 106.

98. "Chova Le-Havtiach Chasit Neota Le-Batim Chadashim," *Davar*, April 28, 1967, 9.

99. Groschka, "Ha-Dira Ba-Aklim Ha-Ham," 82–83.

100. Ernstein, "Mador Le-Chomrei Binyan," *Building in the Near East* 4 (1935): 15.

101. *Mechkarim: Avodot Ha-Mosad Le-Cheker Ha-Bniya Ve-Hatekhnika* (Jerusalem: The Association of Engineers, Architects and Measurers in Eretz Israel, 1944), 2:80.

102. Gordon, "Gagot Mito'fefim Ba-Ruach," 61.

103. Uzi Agassi, *W. J. Wittkower: The Cool Breeze Comes from the West* [Hebrew] (Tel Aviv: Tel Aviv University 1993), 27.

104. Çelik Zeynep, "Le Corbusier, Orientalism, Colonialism," *Assemblage: Critical Journal of Architecture and Design Culture* 17 (April 1992): 61.

105. Ibid., 61–62.

106. *Mehkarim: Avodot Hamosad Leheker Habniya Ve Hatekhnika* (Jerusalem: The Association of Engineers, Architects and Measurers in Eretz Israel, 1944), 1:125.

107. Werner Joseph Wittkower, "Aklim Be-Binyanei Ta'asiya Be-Eretz Israel," *Journal of the Association of Engineers and Architects* 7 (1945): 3.

108. Nigel Whiteley, *Reyner Banham: Historian of the Immediate Future* (Cambridge, MA: MIT Press, 2002), 401.

CHAPTER 4. CLIMATE AND THE STUDY OF PLANTS

1. For example: Ellsworth Huntington, "Climatic Change and Agricultural Exhaustion as Elements in the Fall of Rome," *Quarterly Journal of Economics* 31, no. 2 (February 1917): 173–208.

2. Ellsworth Huntington, *Palestine and Its Transformation* (Boston: Houghton Mifflin Company, 1911), vii.

3. Ibid., 6.

4. J. W. Gregory, "Is the Earth Drying Up?," *The Geographical Journal* 43, no. 2 (February 1914): 152.

5. Ibid., 152–54.

6. David Moon, "The Environmental History of the Russian Steppes: Vasilii Dockuchaev and the Harvest Failure of 1891," *Transactions of the Royal Historical Society* no. 15 (2005): 252–53. According to Richard Grove, the origins of the debate began in the 1860s following the publication of Darwin's controversial theories in *On the Origin of Species by Means of Natural Selection*, which influenced and transformed many fields of thought, and, among other things, stimulated an unprecedented wave of environmental concerns. See *Ecology, Climate and Empire: Colonialism and Global Environmental History, 1400–1940* (Cambridge, UK: White Horse, 1997), 21.

7. "Das Klima Palästinas," *Die Welt*, August 7, 1903, 8.

8. Heinrich Hilderscheid, "Die Niederschlagsverhältnisse Palästinas in alter und neuer Zeit," *Zeitschrift des Deutschen Palästina-Vereins* 25 (1902): 85–86.

9. Dana von Suffrin, *Pflanzen für Palästina: Otto Warburg und die Naturwissenschaften im Jischuw* (Tübingen: Mohr Siebeck, 2019), 70.

10. Ibid., 66; and Aaron Aaronsohn, "The Jewish Agricultural Station and Its Programme," in *Zionist Work in Palestine* ed. Israel Cohen (New York Judean Publishing Company, 1912), 114–21.

11. Diana K. Davis and Edmund Burke III, eds., *Environmental Imaginaries of the Middle East and North Africa* (Athens: Ohio University Press; London, 2011), 2–6. Also see Tamar Novick, *Milk and Honey: Technologies of Plenty in the Making of a Holy Land* (Cambridge, MA: MIT Press, 2023).

12. Caroline Ford, *Natural Interests: The Contest over Environment in Modern France* (Cambridge, MA: Harvard University Press, 2016), 146.

13. It is important to stress, however, that the planting of trees and forests in the first half of the twentieth century was not seen as contributing to a worldwide effort to combat global warming but rather as a local initiative to "fix" and "improve" specific climatic regions (including their "bad air," which allegedly resulted in malaria) that were believed to be degenerating.

14. Irus Braverman, *Planted Flags: Trees, Land, and Law in Israel/Palestine* (Cambridge, UK: Cambridge University Press, 2009), 31–32; Richard Grove, *Green Imperialism: Colonial Expansion, Tropical Island Edens and the Origins of Environmentalism, 1600–1860* (Cambridge, UK: Cambridge University Press, 1995), 45–46; Roza El-Eini, *Mandated Landscape: British Imperial Rule in Palestine, 1929–1948* (London: Routledge, 2006), 190; and James C. Scott, *Seeing Like a State: How Certain Schemes to Improve the Human Condition Have Failed* (New Haven: Yale University Press, 1998), 11–22.

15. Ramachandra Guha and Madhav Gadgil, "State Forestry and Social Conflict in British India," *Past & Present* 123 (1989): 145.

16. Eliezer Livneh, *Aaron Aaronsohn: Ha-Ish Ve-Zmano* (Jerusalem: Bialik, 1969), 47; and Gideon Biger and Nili Liphschitz, "Australian Trees in the Land of Israel 1865-1950," *The Journal of Israeli History* 16, no. 3 (2008): 238. For more on British and Zionist forestation and climate change see David Shorr, "Forest Law in Mandate Palestine," in *Managing the Unknown: Essays on Environmental Ignorance,* ed. Frank Uekotter and Uwe Luebken (New York: Berghahn Books, 2014), 71–90.

17. Another destination for agronomical studies was California, where Asaf Goor and Amihud Goor studied.

18. According to Alon Tal, other forests that were planted by Jews under Rothschild's administration in Palestine were located in Yesod Ha-Ma'ala and Petach Tikwa (*All the Trees of the Forest: Israel's Woodlands from the Bible to the Present* [New Haven: Yale University Press, 2013], 78). See Moshe Smilansky, "Tzel," *Ha-Aretz,* July 7, 1944, 3; and Biger and Liphschitz, "Australian Trees in the Land of Israel 1865-1950," 238. The eucalyptus was discovered in 1777 by James Cook's botanists. In the 1830s there were failed attempts to acclimatize it in Italy. A second attempt to acclimatize it was made in 1852 in French Algeria. The success this time led to the spread of this tree in other Mediterranean areas such as southern France, Spain, Portugal, and eventually also Italy. It was introduced to Palestine in the last decades of the nineteenth century. See Joseph Weitz, *Forest and Afforestation in Israel* (Ramat Gan: Masada, 1970), 67–68.

19. Greg Barton and B. M. Bennett, "Edward Harold Fulcher Swain's Vision of Forest Modernity," *Intellectual History Review* 21, no. 2 (2011): 269; and J. H. Chang and A. D. King, "Towards a Genealogy of Tropical Architecture: Historical Fragments of Power-Knowledge, Built Environment and Climate in the British Colonial Territories," *Singapore Journal of Tropical Geography* 32, no. 3 (2011): 288.

20. Aaron Sandler, "The Health Conditions of Palestine," in *Zionist Work in Palestine,* ed. Israel Cohen (New York: Judaean Publishing Company, 1912), 75–76.

21. Theodor Herzl, "Ha-Yoman 2, 1897–1901," in *Mivchar Kitvei Herzl* (Tel Aviv: M. Neuman, 1934–1948), November 2, 1889, 149.

22. Von Suffrin, *Pflanzen für Palästina,* 51.

23. Max Bodenheimer, "The Jewish National Fund," in *Zionist Work in Palestine,* ed. Israel Cohen (New York: Judaean Publishing Company, 1912), 25.

24. Zvi Shilony, *Jewish National Fund and Settlement in Eretz-Israel 1903–1914* [Hebrew] (Jerusalem: Yad Ben-Zvi, 1990), 73–75.

25. Nathan Bistritzky, *Be-Mesilat Rishonim: Le-Zekher M. Bodenheimer* (Jerusalem: Ha-Keren Ha-Kayemet Le-Israel, 1950–1951), 135. The same idea

is stated in the writings of Akiva Etinger; see *Im Chaklaiim Ivriim Be-Artzenu* (Tel Aviv: Davar, 1945), 54–55.

26. Smilansky, "Tzel."

27. Weitz, *Forest and Afforestation in Israel*, 43.

28. Ibid., 19.

29. Gideon Biger and Nili Liphschitz, *Green Dress for a Country: Afforestation in Eretz Israel: The First Hundred Years 1850–1950* [Hebrew] (Jerusalem: Ariel, 2000), 40–45.

30. Quoted in Alon Tal, *Pollution in a Promised Land: An Environmental History of Israel* (Berkeley: University of California Press, 2002), 80.

31. Joseph Weitz, *Forest Policy in Israel* [Hebrew] (Jerusalem: Jewish National Fund Publication, 1950), 5. This view is also stated by Etinger, *Im Chaklaiim Ivriim Be-Artzenu*, 56.

32. Tal, *Pollution in a Promised Land*, 52–53.

33. Quoted in Zeev Zivan, *From Nitsana to Eilat* [Hebrew] (Sde Boker: Ben-Gurion Research Institute for the Study of Israel and Zionism, 2012), 14.

34. See Yael Zerubavel, *Desert in the Promised Land* (Stanford, CA: Stanford University Press, 2019), 139.

35. Eyal Weizman, *The Conflict Shoreline: Colonization as Climate Change in the Negev Desert* (Göttingen: Steidl, 2015), 30.

36. Novick, *Milk and Honey*, 57.

37. Ibid., 58.

38. Tal, *All the Trees of the Forest*, 32–36; and Biger and Liphschitz, "Australian Trees in the Land of Israel 1865–1950," 240–41.

39. Herbert Samuel, "An Interim Report on the Civil Administration of Palestine, During the Period 1st July, 1920–30th June, 1921," ed. Office of Public Sector Information (London: HMSO, 1921).

40. Shorr, "Forest Law in Mandate Palestine," 75.

41. Greg Barton, *Empire Forestry and the Origins of Environmentalism* (Cambridge, UK: Cambridge University Press, 2002), 7.

42. Shorr, "Forest Law in Mandate Palestine," 78.

43. Israel Gindel, *Imutz Tzmahim* (Tel Aviv: Am Oved, 1956), 36–37; and Biger and Liphschitz, "Australian Trees in the Land of Israel 1865–1950," 236.

44. William Cronon, *Changes in the Land: Indians, Colonists and the Ecology of New England* (New York: Hill and Wang, 1983), 19–20.

45. John H. Elliott, "Colonial Identity in the Atlantic World," in *Colonial Identity in the Atlantic World, 1500–1800*, ed. Nicholas Canny and Anthony Pagden (Princeton: Princeton University Press, 1987), 9–10.

46. Irus Braverman, "Planting the Promised Landscape: Zionism, Nature, and Resistance in Israel/Palestine," *Natural Resources Journal* 49, no. 2 (Spring, 2009): 342–44; Weitz, *Forest and Afforestation in Israel*, 69; Tal, *All the Trees of the Forest*, 79; and Tal, *Pollution in a Promised Land*, 95.

47. Leah Goldberg, "Me-I'nyan Le-I'nyan: Beit Daniel Be-Zikhron Ya'acov," *Al Ha-Mishmar* March 26, 1945, 2.

48. Barton and Bennett, "Edward Harold Fulcher Swain's Vision of Forest Modernity," 276.

49. Letter from Rachel Yanait Ben Zvi to the United States Forest Service in San Francisco, ISA, פ-2078/60.

50. Otto Warburg, "Vegetation in Palestine," in *Zionist Work in Palestine*, ed. Israel Cohen (New York: Judaean Publication Company, 1912), 51. Other similar sources referring to California as a role model for Palestine include Agr. Brill, über die Orangenkultur in Kalifornien, CZA A8/58.

51. Aaron Aaronsohn, "The Jewish Agricultural Station and Its Programme," in *Zionist Work in Palestine*, ed. Israel Cohen (New York: Judean Publishing Company, 1912), 117.

52. Hanan Oppenheimer, *Cultivation of Sub-Tropical and Tropical Fruit Trees* [Hebrew] (The Jewish Agency: Sifriyat Ha-Sade, 1942).

53. "Hashpa'at Ha-Ruchot Ha-Meyabshot Al Atzei Ha-Hadar," *Ha-Sade* 4, no. 12 (1932): 5; and Asaf Grasovski, "Hashpa'at Ha-Hamsinim Al Poryut Etzei Ha-Pri," *Ha-sade* 15 (1936).

54. Weitz, *Forest Policy in Israel*, 11–12; El-Eini, *Mandated Landscape*, 214.

55. Weitz, *Forest and Afforestation in Israel*, 298–99.

56. Ibid., 178.

57. Aaron Aaronsohn, *Agricultural and Botanical Explorations in Palestine* (U.S. Department of Agriculture, Bureau of Plant Industry, Bulletin N180, 1910), 7.

58. Livneh, *Aaron Aaronsohn*, 64, 138, 50,52.

59. Aaronsohn, *Agricultural and Botanical Explorations in Palestine*, 116.

60. Mustafa Kabha and Nahum Karlinsky, *The Lost Orchard: The Palestinian-Arab Citrus Industry, 1850–1950* (Syracuse, NY: Syracuse University Press 2021), 21.

61. Gindel, *Imutz Tzmahim*, 33, 38–40. Other sources which refer to acclimatization during biblical times include for example, Asaf Goor, *Fruits of the Land of Israel* [Hebrew] (Tel Aviv: Chidekel Publishing, 1974).

62. Samer Alatout, "Bringing Abundance into Environmental Politics: Constructing a Zionist Network of Water Abundance, Immigration, and Colonization," *Social Studies of Science* 39, no. 3 (2009): 380.

63. Walter Clay Lowdermilk, *Palestine Land of Promise* (London: Victor Gollancz Ltd, 1944), 35–36.

64. Ibid., 20.

65. Itzhak Elazari-Volkani, *Midot* (Tel Aviv: Ha-Aretz Ve-Ha-Avoda, 1942), 182.

66. Ibid.

67. Timothy Mitchell, *Rule of Experts: Egypt, Techno-Politics, Modernity* (Berkeley, CA: University of California Press, 2002), 26.

68. Alatout, "Bringing Abundance into Environmental Politics," 382–83.

69. Quoted in ibid., 381.

70. Fredrik Meiton, *Electrical Palestine: Capital and Technology from Empire to Nation* (Oakland: University of California Press, 2019), 47.

71. Aaron T. Wolf, *Hydropolitics Along the Jordan River: Scarce Water and Its Impact on the Arab-Israeli Conflict* (Tokyo: United Nations University Press, 1995), 21.

72. Sandra M. Sufian, *Healing the Land and the Nation: Malaria and the Zionist Project in Palestine, 1920–1947* (Chicago: University of Chicago Press, 2007), 136; Amos Nadan, *The Palestinian Peasant Economy under the Mandate: A Story of Colonial Bungling* (Cambridge, MA: Harvard University Press, 2006), 83.

73. Nadan, *The Palestinian Peasant Economy under the Mandate*, 83, 84, 86; Noam G. Seligman, "The Environmental Legacy of the Fallaheen and the Bedouin," in *Between Ruin and Restoration: An Environmental History of Israel*, ed. Daniel E. Orenstein, Char Miller, and Alon Tal (Pittsburgh: University of Pittsburgh Press, 2013), 34.

74. Nadan, *The Palestinian Peasant Economy under the Mandate*, 64–65; Kabha and Karlinsky, *The Lost Orchard*, 50; and Jacob Metzer, *Economics, Land and Nationalism: Issues in Economic History and Political Economy in the Mandate Era and the State of Israel* (Jerusalem: Magnes Press, 2023), 18.

75. Metzer, *Economics, Land and Nationalism*, 18.

76. One of the most extensive and significant upper water development projects in the country began in 1926, with the granting of a seventy-year concession by the British Government to the Zionist Palestine Electric Corporation, founded in 1923 by the Russian-Jewish magnate Pinhas Rutenberg. The main objective of this enterprise was actually to develop a facility at the confluence of the Jordan and Yarmuk rivers for the generation of hydroelectric power. However, developing Palestine's water sources in order to generate electricity would at the same time prepare the ground for large-scale irrigation. Part of Rutenberg's plan was to transform the Lake of Tiberias "into a great natural storage reservoir, providing a steady flow of water in dry seasons as well as in wet" (Sherene Seikaly, *Men of Capital: Scarcity and Economy in Mandate Palestine* [Stanford, CA: Stanford University Press, 2016], 7). See *Power for Palestine: The Rutenberg Concession*, (New York: Zionist Organization of America, 1922), 2.

77. Wolf, *Hydropolitics Along the Jordan River*, 9.

78. Nahum Karlinsky, *California Dreaming: Ideology, Society, and Technology in the Citrus Industry of Palestine, 1890–1939* (Albany: State University of New York Press, 2005), 96–98. Alon Tal claims that between 1924 and 1938,

Zionist colonizing agencies dug 548 wells and almost as many canal systems to tap springs and streams (*Pollution in a Promised Land*, 54).

79. Margreet Zwarteveen, "Transformations to Groundwater Sustainability: From Individuals and Pumps to Communities and Aquifers," *Current Opinion in Environmental Sustainability* 94 (April 2021): 89, 90.

80. Tal, *Pollution in a Promised Land*, 54; Alatout, "Bringing Abundance into Environmental Politics," 374.

81. Alatout, "Bringing Abundance into Environmental Politics," 382–83.

82. Report by Cyril Q. Henriques, 10/07/1939, CZA, A91/5, 2.

83. "Meteorologists at Work," February 13, 1949, CAHUJI, Ashbel Dov, Personal File, 1.

84. Letter from unknown writer to I. Sieff in London, 28/02/1933, CZA, L18/578.

85. Minute of a meeting concerning the Negev in the headquarter offices of the JNF, 12/06/1944, CZA, A246/170, 7.

86. Report of the British Department of Civil Aviation and the Palestine Meteorological Service from 1945, CAHUJI, Meteorology Dep. 1947.

87. *Weather: A Monthly Magazine for All Interested in Meteorology*, published by the authority of the Royal Meteorological Society, July 1947, CAHUJI, Meteorology Dep. 1947.

88. Caroline Elkins and Susan Pedersen, *Settler Colonialism in the Twentieth Century: Projects, Practices, Legacies* (London: Routledge, 2005), 2.

89. Although Zionism is often portrayed as a socialist movement, recent studies have increasingly shown its concurrent capitalist ambitions since the Mandate period—mainly in the agricultural sector. The primary objective of the Zionist movement—the creation of a stable and sovereign Jewish majority in Palestine—was understood as something that could only be achieved on the basis of a strong economic system that would produce goods and sell them for profit, thus naturally abandoning agricultural models based on production for self-consumption and relying on an industrial economy integrated in the global market. In this respect, Zionist agriculture was capitalist from its beginning. For further reading on this subject see, for example, Matan Kaminer, "Towards a Political Ecology of Zionism in the Rural Sphere," [Hebrew] *Teorya U-Viķoret (Theory and Criticism)* 57 (Winter 2023); Nahum Karlinsky, *Citrus Blossoms: Jewish Entrepreneurship in Palestine, 1890–1939*[Hebrew] (Jerusalem: Magnes, 2000); Meiton, *Electrical Palestine*, 9; and Yael Raviv, *Falafel Nation : Cuisine and the Making of National Identity in Israel* (Lincoln: University of Nebraska Press, 2015). Such aspirations can also be seen in primary sources, such as in report entitled *Die Industrialisierung Palästina* from March 1931, CZA A112/11/1.

90. Shmuel Stoller, *Edut: Levatim Be-Hakamat Meshek Shlechin Be-Bika'at Kineret* (Tel Aviv: Hotza'at Ha-Merkaz Ha-Chaklai, 1980), 13; and Tal, *Pollution in a Promised Land*, 50.

91. Hershkowitz, "Hashka'at Mata'ei Gefen Ve-Atzei Pri Be-Midronei He-Harim," *Ha-Sadeh*, April 1955, 2, located in PLILMR, IV-293–45.

92. Another reason for the Jewish rejection of this crop was related to the consumption habits of the Jewish settler society, as well as of the European markets that were targeted for the export economy. Hershkowitz suggested that olives were not very popular among northern European consumers, who were "suspicious of this fruit and are not used to eating it" (ibid., PLILMR, IV-293–44).

93. Today, as a result of rising temperatures in the region, many "European" fruit crops are suffering from severe damage due to the insufficient number of cold days. See:Lee Yaron, "'The Winter Used to be in November, Now It's in January': Peaches and Apricots Are Disappearing and This is Only the Beginning," *Ha-Aretz*, January 18, 2022, https://www.haaretz.co.il/nature/climate/2022-01–18/ty-article-magazine/.highlight/0000017f-e702-d62c-a1ff-ff7bc0f70000.

94. Oppenheimer, *Cultivation of Sub-Tropical and Tropical Fruit Trees*; Report by Hershkowitz Elhanan, "Hatza'ot Ha-Ve'ada Le-Gidul Atzei Pri Sub-Tropiim" (Suggestions of the Committee for Sub-tropic Fruit Cultivation), 06/07/1944, PLILMR, IV 293–75; and "150 Varieties of Fruit," *Palestine Post Jerusalem*, 16.06.1933, CZA S90/2266/40. The British government similarly examined the acclimatization of foreign fruits in its horticultural stations. These examinations included the planting of, for instance, avocado, raspberry, gooseberry, blackberry, lichee, pawpaw, annona, mango, and guava (El-Eini, *Mandated Landscape*, 127). Some of these fruits had already been cultivated in the region for centuries and even millennia. For more see Sureshkumar Muthukumaran, *The Tropical Turn: Agricultural Innovation in the Ancient Middle East and the Mediterranean* (Berkeley: University of California Press, 2023).

95. Water quantities depended on the size of trees, type of soil, and specific climate conditions where they grew. Nevertheless, a mature orange tree generally uses around 17 gallons of water per day in the winter and 135 gallons of water per day in summer. See Glenn C. Wright, "Irrigating Citrus Trees" (February 2000), 1.

96. See Israel Gindel, *The Coffee Plant and Its Growth in Israel* [Hebrew] (Tel Aviv: Sifriyat Ha-Sade, 1959), 47.

97. The Cavendish banana (which was also called *Masri* in Palestine) is believed to have travelled from South China to Mauritius around the 1830s and via the English botanical gardens to the Canary Islands. In the late nineteenth century, it was brought to Egypt and from there to Palestine (Goor, *Fruits of the Land of Israel*, 320–22). See PLILMR, IV-293–8; and see Stoller, *Edut*, 29–30.

98. Goor, *Fruits of the Land of Israel*, 333.

99. Stoller, *Edut*, 27–28.

100. Report of the Fruit Growers' Union, 13–14/03/1944, PLILMR, IV-293–75.

101. Tomato plantations were also "suffering" from too much heat in this region and after several experiments, it was decided to cover their soil to keep its temperature from rising too high during the summer days (Stoller, *Edut*, 65–66).

102. George M. Odlum, "Gidul Ha-Bananot Be-Palestina" (n.p., Department of Agriculture and Forestry, 1927), 10–11.

103. See, for example, a lecture by Dr. Manzikovski, Director of the Chemical Department in the Agricultural Experimentation Station in Tel Aviv, unknown date, in PLILMR, IV-293- 8.

104. The same is said about irrigation attempts in the Beisan Valley in the same report (Anti-Malaria Measures—Reports of Investigations, Malaria Survey Unit, ISA, מ/32/654).

105. A Study on the Cultivation of Bananas in the Canary Islands, PLILMR, IV-293–8.

106. Nadan, *The Palestinian Peasant Economy under the Mandate*, 63–64.

107. Diana K. Davis, *The Arid Lands: History, Power, Knowledge* (Cambridge, MA: The MIT Press, 2016), 5.

108. Alon Tal, "To Make a Desert Bloom: The Israeli Agricultural Adventure and the Quest for Sustainability," *Agricultural History* 81, no. 2 (2007).

109. El-Eini, *Mandated Landscape*, 121; Karlinsky, *California Dreaming*, 13, 92; Karlinsky, *Citrus Blossoms*, 10.

110. Stoller, *Edut*, 16.

111. Karlinsky, *California Dreaming*.

112. Warburg, "Vegetation in Palestine," 51.

113. Palestine's desired export economy was also related to improvements in transportation and the reduced cost of moving food. Palestine's location on the world's main trade routes—on the coast of the Mediterranean, between Asia and Africa, next to the Suez Canal and the Haifa–Dera'a railway—was central to deciding on its agricultural-economic strategy.

114. Karlinsky, *Citrus Blossoms*, 17; and Akiva Etinger "Mediniyut Ha-Tsuna Be-Angliya U-Be-Germanya" (Nutrition Policy in England and Germany), unknown date, 3–4, CZA, A111/25. Although Britain was one of the main importers of fruits from Palestine, Mandate Palestine did not profit from any Imperial tax exemptions or benefits. In addition, bananas also became popular because of their hygienic value. According to Virginia Scott Jenkins, this fruit was recommended by health experts because its peel was seen as a "germ-proof wrapper" (*Bananas: An American History* [Washington, DC: Smithsonian Institution Press, 2000]). See David Grigg, "The Changing Geography of World Food Consumption in the Second Half of the Twentieth Century," *The Geographical Journal* 165, no. 1 (March 1999): 1; and Metzer, *Economics, Land and Nationalism*, 101.

115. "Palestine as Europe's Granary: Interesting Prediction by British Food Expert," *Palestine Post*, September 21, 1942, 9, CZA A111/67.

116. Odlum, "Gidul Ha-Bananot Be-Palestina," 9.

117. S. E. Soskin, *Small Holding and Irrigation: The New Form of Settlement in Palestine* (London: Jewish National Fund, 1920), 19.

118. Report by Hershkowitz Elhanan, "Hatza'ot Ha-Veada Le-Gidul Atzei Pri Sub-Tropiim" (Suggestions of the Committee for Sub-tropic Fruit Cultivation), July 6, 1944, PLILMR, IV-293–75.

119. Eulogy Booklet for Elhanan Hershkowitz, 1982, PLILMR, IV-293–44.

120. Ibid.

121. Akiva Etinger, "Ma Hem Ha-Mitzrakhim She-Chaklautenu Ma'amida Le-Reshut Ha-Isha?", *Hakol La-Isha,* unkonwn date, CZA, A111/25.

122. "Fruit for Health," *The Palestine Post,* February 8, 1939, 6.

123. Efrat Gilad, "Meat in the Heat: A History of Tel Aviv under the British Mandate for Palestine (1920s-1940s)" (PhD diss., University of Geneva, 2021), 93–95. According to Sherene Seikaly a direct trade existed between Palestinian farmers and the Jewish markets throughout the 1930s, even during the 1936 Strike. Especially in Haifa and Acre, many Jews shopped in Arab markets where the prices were usually lower (Yishuv produce was on average 21 percent more expensive than Palestinian produce before the war) (*Men of Capital,* 132).

CONCLUSIONS

1. See, for example, Jason R. Rohr, "Frontiers in Climate Change–Disease Research," *Trends in Ecology & Evolution* 26, no. 6 (2011): 270–77; J. Luck, "Climate Change and Diseases of Food Crops," *Plant Pathology* 60, no. 1 (2011): 113–21; Kevin D. Lafferty, "The Ecology of Climate Change and Infectious Diseases," *Ecology* 90, no. 4 (2009): 888–900; and Anil Kumar Misra, "Climate Change and Challenges of Water and Food Security," *International Journal of Sustainable Built Environment* 3, no. 1 (2014): 153–65.

2. A similar claim was recently made by Matan Kaminer, "Towards a Political Ecology of Zionism in the Rural Sphere," *Teorya U-Viḳoret (Theory and Criticism)* 57 (Winter 2023): 71–99.

3. Amos Kenan, *Your Land, Your Country [Hebrew]* (Tel Aviv: Yediot Aharonot, 1981), 9.

4. Quoted in Netta Ahituv, "The Data Is Irrefutable: Israeli Winter Is Going Extinct," *Ha'Aretz,* December 23, 2022, https://www.haaretz.com/israel-news/2022-12-23/ty-article-magazine/.highlight/the-data-is-irrefutable-israeli-winter-is-going-extinct/00000185-3f8b-dc10-a7d7-7fffd13a0000.

5. Matan Kaminer, Basma Fahoum, and Edo Konrad, "From Heat Waves to 'Eco-apartheid': Climate Change in Israel-Palestine," *+972 Magazine,* August 8, 2019, https://www.972mag.com/climate-change-israel-palestine.

6. Diana K. Davis, *The Arid Lands: History, Power, Knowledge* (Cambridge, MA: MIT Press, 2016), 6–7.

7. Cited in Andrea Smith, "Queer Theory and Native Studies the Heteronormativity of Settler Colonialism," *GLQ* 16, no. 1–2 (2010): 53.

8. Important recent research on the Negev has not specifically addressed these issues; see Eyal Weizman, *The Conflict Shoreline: Colonization as Climate Change in the Negev Desert* (Göttingen: Steidl, 2015); Yael Zerubavel, *Desert in the Promised Land* (Stanford, CA: Stanford University Press, 2019).

Bibliography

ARCHIVAL SOURCES

BAA   Beit Aaronsohn Archive, Zikhron Ya'akov
CAHUJI   The Central Archives of the Hebrew University of Jerusalem,
           Jerusalem
CZA   The Central Zionist Archives, Jerusalem
AGSJI   The Archives of German-Speaking Jewry in Israel, Tefen/Haifa
HHA   Haganah Historical Archives, Tel Aviv
ISA   Israel State Archives, Jerusalem
LBIJER   Leo Baeck Institute, Jerusalem
LBICJH   Leo Baeck Institute in the Center for Jewish History, New York City
TNA   The National Archives, London
PLILMR   Pinchas Lavon Institute for Labour Movement Research, Tel Aviv

PRINTED SOURCES

Aaronsohn, Aaron. *Agricultural and Botanical Explorations in Palestine.* U.S.
    Department of Agriculture, Bureau of Plant Industry, Bulletin N180, 1910.
———. "The Jewish Agricultural Station and Its Programme." In *Zionist Work
    in Palestine,* edited by Israel Cohen, 114–21. New York: Judean Publishing
    Company, 1912.

Agassi, Uzi. *W. J. Wittkower: The Cool Breeze Comes from the West* [Hebrew]. Tel Aviv: Tel Aviv University 1993.

Agnon, Shmuel Yosef. *Only Yesterday.* Princeton, NJ: Princeton University Press, 2000 [1945].

Ahituv, Netta. "The Data Is Irrefutable: Israeli Winter Is Going Extinct." *Ha'Aretz*, December 23, 2022. https://www.haaretz.com/israel-news /2022-12-23/ty-article-magazine/.highlight/the-data-is-irrefutable-israeli-winter-is-going-extinct/00000185-3f8b-dc10-a7d7-7fffd13a0000.

Alatout, Samer. "Bringing Abundance into Environmental Politics: Constructing a Zionist Network of Water Abundance, Immigration, and Colonization." *Social Studies of Science* 39, no. 3 (2009): 363–94.

Aleksandrowicz, Or. "Appearance and Performance: Israeli Building Climatology and Its Effect on Local Architectural Practice (1940–1977)." *Architectural Science Review* 60, no. 5 (2017): 371–81.

———. *Daring the Shutter: The Tel Aviv Idiom of Solar Protections.* Tel Aviv: Public School, 2015.

———. "Kurkar, Melet, Aravim, Jehudim: Eikh Bonim Ir Ivrit." *Teorya U-Vikoret (Theory and Criticism)* 36 (2010): 61–87.

———. "Mishelet Churban: Chayav U-Moto Shel Binyan Ha-Gymnasia Hertseliya." In *Yemei Ha-Gymnasia*, edited by Guy Raz, 26–47. Tel Aviv: Eretz Israel Museum, 2013.

Allweil, Asher. "Nituach Tzurot Ha-Binuy Ve-Shitot Ha-Bitzua Be-Mifa'lei Ha-Shikun La-Olim." *Journal of the Association of Engineers and Architects* 9 (1950): 2–15.

Almog, Oz. *The Sabra—A Profile* [Hebrew]. Tel Aviv: Am Oved, 1997.

Alroey, Gur. "Journey to New Palestine: The Zionist Expedition to East Africa and the Aftermath of the Uganda Debate." *Jewish Culture and History* 10, no. 1 (May 2012): 23–58.

———. *An Unpromising Land: Jewish Migration to Palestine in the Early Twentieth Century.* Stanford, CA: Stanford University Press, 2014.

Alterman, Natan. "Shir Boker." In *Pizmonim Ve-Shirei Zemer*, 2:302–3. Tel Aviv: Ha-Kibbutz Ha-Meuchad, 1979 [1934].

Anderson, Warwick. *The Cultivation of Whiteness: Science, Health and Racial Destiny in Australia.* Carlton, Australia: Melbourne University Press, 2005.

Arnold, David. *Imperial Medicine and Indigenous Societies.* Manchester, UK: Manchester University Press, 1988.

———. *Warm Climates and Western Medicine: The Emergence of Tropical Medicine, 1500–1900.* Amsterdam: Rodopi, 1996.

Ashbel, Dov. *Avir Artzenu.* Tel Aviv: Omanut Ba'am, 1940.

Auerbach, Elias. *Palästina Als Judenland.* Berlin: Aktionskomitee der zionistischen Organisation, in Kommission beim Juedischen Verlag, 1912.

Awin, Josef. "Al Omanut Ha-Bniya Be-Eretz Israel." *Binyan Ve-Charoshet* 6, no. 3–4 (1927): 12.

B. S. "Gam Elle Goalim." *He-Ahdut*, July 10, 1914.

Banivanua-Mar, Tracey, and Penelope Edmonds. *Making Settler Colonial Space: Perspectives on Race, Place and Identity.* Basingstoke, UK: Palgrave Macmillan, 2010.

Bar-Yosef, Eitan. *The Holy Land in English Culture 1799–1917: Palestine and the Question of Orientalism.* Oxford: Clarendon, 2005.

———. "Spying out the Land: The Zionist Expedition to East Africa, 1905." In *'The Jew' in Late-Victorian and Edwardian Culture: Between the East End and East Africa*, edited by Eitan Bar-Yosef and Nadia Valman, 183–200. Basingstoke, UK: Palgrave Macmillan, 2009.

Barak, On. *Powering Empire: How Coal Made the Middle East and Sparked Global Carbonization.* Oakland: University of California Press, 2020.

Barton, Greg. *Empire Forestry and the Origins of Environmentalism.* Cambridge, UK: Cambridge University Press, 2002.

Barton, Greg, and B. M. Bennett. "Edward Harold Fulcher Swain's Vision of Forest Modernity." *Intellectual History Review* 21, no. 2 (2011): 135–50.

Beckmann, D. "Eikh Le-Hishamer Mi-Malaria." In *Briut Ha-Oleh: Madrikh La-Oleh Ha-Chadash Be-Inyanei Higyena U-Vriut*. 13–15. Jerusalem: Kupat Cholim Amamit Ve-Machleket Ha-Aliya Shel Ha-Sokhnut Ha-Yehudit, 1939.

"Beit Ha-Charoshet Le-Kerach Tel Aviv." *Ha-Aretz*, May 3, 1928.

Ben-Amos, Avner, and Ofri Ilany. *Sacred People: Bible and Nationalism in the Modern Era* [Hebrew]. Jerusalem: Magnes Press, 2021.

Ben Zeev, Nimrod. "Foundations of Inequality: Construction, Political Economy, Race and the Body in Palestine/Israel, 1918–1973." PhD diss., University of Pennsylvania, 2020.

Beriut Ha-Oleh: Madrich La-Oleh He-Hadash Be-Inyanei Higyena U-Vriut. Jerusalem: Kupat Holim Amamit Ve-Mahleket Ha-Aliya Shel Ha-Sokhnut Ha-Yehudit, 1939.

Berwald, Alexander. "Bauliche Probleme in Palästina." *Die Welt* 41 (1910): 1046–49.

Bhabha, Homi K. "The White Stuff (Political Aspect of Whiteness)." *Art Forum* 36, no. 9 (1998): 21–24.

Biger, Gideon, and Nili Liphschitz. "Australian Trees in the Land of Israel 1865–1950." *The Journal of Israeli History* 16, no. 3 (2008): 235–44.

———. *Green Dress for a Country: Afforestation in Eretz Israel: The First Hundred Years 1850–1950* [Hebrew]. Jerusalem: Ariel, 2000.

Binyamini, Dr. A. "Ha-Bechinot U-Veriut Yeladeinu." *Ha-Aretz*, July 24, 1930, 3.

Bistritzky, Nathan. *Be-Mesilat Rishonim: Le-Zekher M. Bodenheimer.* Jerusalem: Ha-Keren Ha-Kayemet Le-Israel, 1950–1951.

Blanckenhorn, Max. "Studien über das Klima des Jordantals." *Zeitschrift des Deutschen Palästina-Vereins* 32, no. 1/2 (1909): 38–109.

———. "Zum Klima von Palästina." *Zeitschrift des Deutschen Palästina-Vereins* 33, no. 2/3 (1910): 107–64.

Bloom, Etan. *Arthur Ruppin and the Production of Pre-Israeli Culture*. Leiden: Brill; Biggleswade, 2011.

Bodenheimer, Max. "The Jewish National Fund." In *Zionist Work in Palestine*, edited by Israel Cohen, 25–29. New York: Judaean Publishing Company, 1912.

Boyarin, Daniel. *Unheroic Conduct: The Rise of Heterosexuality and the Invention of the Jewish Man*. Berkeley: University of California Press, 1997.

Boym, Svetlana. *The Future of Nostalgia*. New York: Basic Books, 2001.

Braverman, Irus. *Planted Flags: Trees, Land, and Law in Israel/Palestine*. Cambridge, UK: Cambridge University Press, 2009.

———. "Planting the Promised Landscape: Zionism, Nature, and Resistance in Israel/Palestine." *Natural Resources Journal* 49, no. 2 (Spring 2009): 317–65.

———. *Settling Nature: The Conservation Regime in Palestine-Israel*. Minneapolis: University of Minnesota Press, 2023.

Briggs, John, Joanne Sharp, Hoda Yacoub, Nabila Hamed, and Alan Roe. "The Nature of Indigenous Environmental Knowledge Production: Evidence from Bedouin Communities in Southern Egypt." *Journal of International Development* 19 (2007): 239–51.

Buckle, Henry Thomas. *History of Civilisation in England*. London: John W. Parker & Son, 1857.

Busbridge, Rachel. "Israel-Palestine and the Settler Colonial 'Turn': From Interpretation to Decolonization." *Theory, Culture & Society* 35, no. 1 (2018): 91–115.

Canaan, Tawfik. "Aberglaube und Volksmedizin im Lande der Bibel." *Abhandlungen des hamburgischen Kolonialinstituts* 20, no. 3 (1914).

———. "Der Kalender des palästinensischen Fellachen." *Zeitschrift des Deutschen Palästina-Vereins* 36, no. 4 (1913): 266–300.

———. "Folklore of the Seasons in Palestine." *Journal of the Palestine Oriental Society* 3 (1923): 21–35.

———. *The Palestinian Arab House, Its Architecture and Folklore*. Jerusalem: Syrian Orphanage Press, 1933.

Chang, J. H., and A. D. King. "Towards a Genealogy of Tropical Architecture: Historical Fragments of Power-Knowledge, Built Environment and Climate in the British Colonial Territories." *Singapore Journal of Tropical Geography* 32, no. 3 (2011): 283–300.

"Chova Le-Havtiach Chazit Neota Le-Batim Chadashim." *Davar*, April 28, 1967, 9.

Coen, Deborah R. "Climate and Circulation in Imperial Austria." *The Journal of Modern History* 82, no. 4 (December, 2010): 839–75.

Cohen, Netta. "Environment, Climate and the East in the Eyes of Jewish Architects in Palestine 1909–1948." MA thesis, Tel Aviv University, 2013.

———. "Shades of White: African Climate and Jewish European Bodies, 1903–1905." *The Journal of Imperial and Commonwealth History* 50, no. 2 (2022): 298–316.

Cooper, Gail. *Air-Conditioning America: Engineers and the Controlled Environment, 1900–1960.* Baltimore: Johns Hopkins University Press, 1998.

Crinson, Mark. *Modern Architecture and the End of Empire.* Aldershot, UK: Ashgate, 2003.

Cronon, William. *Changes in the Land: Indians, Colonists and the Ecology of New England.* New York: Hill and Wang, 1983.

Crozier, Anna. "Sensationalising Africa: British Medical Impressions of Sub-Sahara Africa, 1890–1939." *Journal for Imperial and Commonwealth History* 35, no. 3 (2007): 393–415.

Davis, Diana K. *The Arid Lands: History, Power, Knowledge.* Cambridge, MA: The MIT Press, 2016.

———. *Resurrecting the Granary of Rome: Environmental History and French Colonial Expansion in North Africa.* Athens: Ohio University Press, 2007.

Davis, Diana K., and Edmund Burke III, eds. *Environmental Imaginaries of the Middle East and North Africa.* Athens: Ohio University Press, 2011.

Degani, Arnon. "On the Frontier of Integration: The Histadrut and the Palestinian Arab Citizens of Israel." *Middle Eastern Studies* 56, no. 3 (2020): 412–16.

Duany, Maya. *Ha-Agam Ha-Ne'elam: Ha-Chula Khe-Semel Le-Hishtanut Ha-Yachas La-Sviva Be-Medinat Israel.* Jerusalem: Yad Ben-Zvi, 2020.

Duempelmann, Sonja. *Seeing Trees: A History of Street Trees in New York City and Berlin.* New Haven: Yale University Press 2019.

Efrat, Zvi. *The Israeli Project, Building and Architecture 1948–1973* [Hebrew]. Tel Aviv: Tel Aviv Museum of Art, 2004.

Efron, John M. *Defenders of the Race: Jewish Doctors and Race Science in Fin-De-Siecle Europe.* New Haven: Yale University Press, 1994.

El-Eini, Roza. *Mandated Landscape: British Imperial Rule in Palestine, 1929–1948.* London: Routledge, 2006.

Elazari-Volkani, Itzhak. *Midot.* Tel Aviv: Ha-Aretz Ve-Ha-Avoda, 1942.

"Elef Tzarkhanei Kerach Be-Tel Aviv." *Ha-Mashkif,* June 10, 1946, 40.

Elhanani, Avraham Haim. "Bniya Ve-Tikhnun Lefi Aklim Ha-Aretz." *Davar,* July 20, 1949, 8.

Elkins, Caroline, and Susan Pedersen. *Settler Colonialism in the Twentieth Century : Projects, Practices, Legacies.* London: Routledge, 2005.

Elliott, John H. "Colonial Identity in the Atlantic World." In *Colonial Identity in the Atlantic World, 1500–1800,* edited by Nicholas Canny and Anthony Pagden, 3–14. Princeton, NJ: Princeton University Press, 1987.

Endelman, Todd. "Anglo-Jewish Scientists and the Science of Race." *Jewish Social Studies* 11, no. 1 (Autumn 2004): 52–92.

Ernstein. "Mador Le-Chomrei Binyan." *Building in the Near East* 4 (1935): 15.

Etinger, Akiva. *Im Chaklaiim Ivriim Be-Artzenu.* Tel Aviv: Davar, 1945.

Fahmy, Khaled. "An Olfactory Tale of Two Cities: Cairo in the Nineteenth Century." In *Historians in Cairo: Essays in Honor of George Scanlon,* edited by Jill Edwards, 155–87. Cairo: American University in Cairo Press, 2002.

Falk, Raphael. *Zionism and the Biology of the Jews.* New York: Springer, 2017.

———. "Zionism and the Biology of the Jews." *Science in Context* 11, no. 3–4 (1998): 587–607.

Farley, John. *Bilharzia: A History of Imperial Tropical Medicine.* Cambridge, UK: Cambridge University Press, 1991.

Feige, Rudolf. "He'arot Al Ha-Klimatologia Ha-Shimushit Be-Eretz Israel." In *Ha-Adam Veha-Aklim Be-Eretz Israel,* edited by S. Rosenbaum, 10–40. Tel Aviv: Olympia, n.d.

Fenster, Tovi. *Whose City Is It? Planning, Knowledge and Everyday Life* [Hebrew]. Tel Aviv: Ha-Kibbutz Ha-Meuchad, 2012.

Fletcher, Robert. *British Imperialism and 'the Tribal Question': Desert Administration and Nomadic Societies in the Middle East, 1919–1936.* Oxford: Oxford University Press, 2015.

Ford, Caroline. *Natural Interests: The Contest over Environment in Modern France.* Cambridge, MA: Harvard University Press, 2016.

———. "Nature, Culture and Conservation in France and Her Colonies 1840–1940." *Past & Present* 183 (2004): 173–98.

Fritzsche, Peter. *A Nation of Fliers: German Aviation and the Popular Imagination.* Cambridge, MA: Harvard University Press, 1992.

Furani, Khaled, and Dan Rabinowitz. "The Ethnographic Arriving of Palestine." *Annual Review of Anthropology* 40 (2011): 475–91.

Geddes, Patrick. *Town-Planning Report—Jaffa and Tel Aviv.* N.p., 1925.

Gibbons, Major A. St. Hill, Alfred Kaiser, and Nahum Wilbush. "Report on the Work of the Commission Sent out by the Zionist Organization to Examine the Territory Offered by H. M. Government to the Organization for the Purpose of a Jewish Settlement in British East Africa." London : Wertheimer, Lea & Co, May 1905.

Gilad, Efrat. "Camel Controversies and Pork Politics in British Mandate Palestine." *Global Food History* (2022): https://doi.org/10.1080/20549547.2022.2106074.

———. "Meat in the Heat: A History of Tel Aviv under the British Mandate for Palestine (1920s–1940s)." PhD diss., University of Geneva, 2021.

Gilman, Sander L. *The Jew's Body*. London: Routledge, 1991.

Gindel, Israel. *The Coffee Plant and Its Growth in Israel* [Hebrew]. Tel Aviv: Sifriyat Ha-Sade, 1959.

———. *Imutz Tzmahim*. Tel Aviv: Am Oved, 1956.

Ginzburg, Shlomo. "Ha-Orientatzya Shel Ha-Dira Be-Haifa." *Building in the Near East* 9–10 (November 1936): 6.

Goldberg, Leah. "Me-I'nyan Le-I'nyan: Beit Daniel Be-Zikhron Ya'acov." *Al Ha-Mishmar*, March 26, 1945.

Goor, Asaf. *Fruits of the Land of Israel*. [Hebrew]. Tel Aviv: Chidekel Publishing, 1974.

Gordon, Gil. "Gagot Mito'fefim Baruach: Knisatam Shel Re'afim Ve-Ta'asiyat Ha-Charsit Le-Eretz Israel." *Zmanim: A Historical Quarterly* 96 (Spring 2006): 58–67.

Grasovski, Asaf. "Hashpa'at Ha-Hamsinim Al Poryut Etzei Ha-Pri." *Ha-Sade* 15 (1936): 222–25.

Green, Rayna. "The Tribe Called Wannabee: Playing Indian in America and Europe." *Folklore* 99, no. 1 (1988): 30–55.

Greenberg, Zalman, and Anton Alexander. "Israel Jacob Kligler: The Story of a 'Little Big Man,' a Giant in the Field of Public Health in Palestine." *Korot* 21 (2011–2012): 175–206.

Gregory, J. W. "Is the Earth Drying Up?" *The Geographical Journal* 43, no. 2 (Feb. 1914): 148–72.

Grigg, David. "The Changing Geography of World Food Consumption in the Second Half of the Twentieth Century." *The Geographical Journal* 165, no. 1 (March 1999): 1–11.

Dr. Groschka. "Ha-Dira Ba-Aklim Ha-Ham." In *Ha-Adam Ve-Ha-Aklim Be-Eretz Israel*, edited by S. Rosenbaum, 67–86. Tel Aviv: Olympia, n.d.

Grove, Richard. *Ecology, Climate and Empire: Colonialism and Global Environmental History, 1400–1940*. Cambridge, UK: White Horse, 1997.

———. *Green Imperialism: Colonial Expansion, Tropical Island Edens and the Origins of Environmentalism, 1600–1860*. Cambridge, UK: Cambridge University Press, 1995.

Guha, Ramachandra, and Madhav Gadgil. "State Forestry and Social Conflict in British India." *Past & Present* 123 (1989): 141–77.

Gurevitch, Zali, and Gideon Aran. "On Israeli and Jewish Place: Israeli Anthropology." In *On Israeli and Jewish Place* [Hebrew], edited by Zali Gurevitch, 22–74. Tel Aviv: Am Oved, 2007.

Gutman, Nahum. "Mashehu Al Tzurata Ha-Chitztonit Shel Ha-Ir." In *Tel Aviv—a Silver Jubilee: A Documentary Anthology* [Hebrew], edited by Maoz Azaryahu, Arnon Golan and Aminadav Dikman, 129. Jerusalem: Carmel Publishing, 2009 [1933].

Ha-levi, Abraham Jehuda. *Briut Ha-Tzibur*. Tel Aviv: Achiever, 1935.

"Haifa Shuv Sha'aruruyat Kerach." *Ha-Tzofe*, August, 29, 1945, 2.

Harper, Kristine. *Weather by the Numbers: The Genesis of Modern Meteorology.* Cambridge, MA: MIT, 2008.

Harrison, Mark. *Climates & Constitutions: Health, Race, Environment and British Imperialism in India, 1600–1850.* Oxford: Oxford University Press, 1999.

———. "'The Tender Frame of Man': Disease, Climate, and Racial Difference in India and the West Indies, 1760–1860." *Bulletin of the History of Medicine* 70 (1996): 68–93.

Hart, Mitchell Bryan. *Social Science and the Politics of Modern Jewish Identity.* Stanford, CA: Stanford University Press, 2000.

"Hashpa'at Ha-Ruchot Ha-Meyabshot Al Atzei Ha-Hadar." *Ha-sade* 4, no. 12 (1932): 152–56.

Helman, Anat. *Urban Culture in 1920s and 1930s Tel Aviv* [Hebrew]. Haifa: Haifa University Press, 2007.

———. *Young Tel Aviv: A Tale of Two Cities.* Hanover, NH: University Press of New England, 2010.

Herzl, Theodor. *Altneuland = Old-New Land: Novel.* Haifa: Haifa Pub. Co, 1960 [1902].

———. *The Complete Diaries of Theodor Herzl.* Edited by Raphael Patai. Vol. I. New York: Herzl Press, 1961.

———. "Ha-Yoman 2, 1897–1901." In *Mivchar Kitvei Herzl.* Tel Aviv: M. Neuman, 1934–1948.

Hilderscheid, Heinrich. "Die Niederschlagsverhältnisse Palästinas in alter und neuer Zeit." *Zeitschrift des Deutschen Palästina-Vereins* 25 (1902): 5–105.

Hirsch, Dafna. *"We Are Here to Bring the West": Hygiene Education and Culture Building in the Jewish Society of Palestine During the British Mandate Period.* [Hebrew]. Sde Boker: The Ben-Gurion Research Institute for the Study of Israel and Zionism, 2014.

———. "'We Are Here to Bring the West, Not Only to Ourselves': Zionist Occidentalism and the Discourse of Hygiene in Mandate Palestine." *International Journal of Middle East Studies* 41, no. 4 (2009).

———. "Zionist Eugenics, Mixed Marriage, and the Creation of a 'New Jewish Type'." *The Journal of the Royal Anthropological Institute* 15, no. 3 (2009): 592–609.

Hirsch, Luise. *From the Shtetl to the Lecture Hall: Jewish Women and Cultural Exchange.* Lanham, MD: University Press of America, 2013.

Home, Robert. *On Planting and Planning: The Making of British Colonial Cities.* London: Taylor & Francis, 1997.

Howard, Sir Ebenezer. *Garden Cities of To-Morrow.* London: Swan Sonnenschein & Co., 1902.

Hulme, Mike. "The Conquering of Climate: Discourses of Fear and Their Dissolution." *The Geographical Journal* 174, no. 1 (March 2008): 5–16.

Huntington, Ellsworth. "Climatic Change and Agricultural Exhaustion as Elements in the Fall of Rome." *Quarterly Journal of Economics* 31, no. 2 (February, 1917): 173–208.

———. *Palestine and Its Transformation*. Boston: Houghton Mifflin Company, 1911.

Hysler-Rubin, Noah. *Patrick Geddes and Town Planning: A Critical View.* London: Routledge, 2011.

———. "A Cultural Acropolis in Tel Aviv: Patrick Geddes' Vision for the First Hebrew City" [Hebrew]. *Zmanim: A Historical Quarterly* 106 (Spring, 2009): 36–40.

Inal, Onur, and Yavuz Kose. *Seeds of Power: Explorations in Ottoman Environmental History*. Winwick, UK: The White Horse Press, 2019.

"Israel Meyatzet Memazgei Avir." *Shearim*, July 13, 1956, 8.

Jabotinsky, Vladimir Zeev. *Ktavim Tziyoniyim Rishonim*. Jerusalem: A. Jabotinsky,1905.

Jenkins, Virginia Scott. *Bananas: An American History*. Washington, DC: Smithsonian Institution Press, 2000.

Jennings, Eric T. *Curing the Colonizers: Hydrotherapy, Climatology, and French Colonial Spas*. Durham: Duke University Press, 2006.

Johnson, Ryan. "Networks of Imperial Tropical Medicine: Ideas and Practices of Health and Hygiene in the British Empire, 1895–1914." PhD diss., University of Oxford, 2009.

"Das jüedische Flugwesen in Eretz Israel." *Ha-Masad* 4, no. 12 (1939).

Kabha, Mustafa, and Nahum Karlinsky. *The Lost Orchard: The Palestinian-Arab Citrus Industry, 1850–1950*. Syracuse, NY: Syracuse University Press 2021.

Kahana, Freddy. *Neither Town nor Village—the Architecture of the Kibbutz 1910–1990*. [Hebrew]. Rishon Le-Tzion: Ha-Moatza Le-Shimur Atarim, 2011.

Kaminer, Matan. "Towards a Political Ecology of Zionism in the Rural Sphere" [Hebrew]. *Teorya U-Viḳoret* (Theory and Criticism) 57 (Winter 2023): 71–99.

Kaminer, Matan, Basma Fahoum, and Edo Konrad. "From Heat Waves to 'Eco-apartheid': Climate Change in Israel-Palestine." *+972 Magazine*, August 8, 2019. https://www.972mag.com/climate-change-israel-palestine.

Kamminga, Harmke, and Andrew Cunningham. *The Science and Culture of Nutrition, 1840–1940*. Amsterdam: Rodopi, 1995.

Kapil, Raj. "Colonial Encounters and the Forging of New Knowledge and National Identities: Great Britain and India, 1760–1850." *Osiris* 15 (2000): 116–34.

Karlinsky, Nahum. *California Dreaming: Ideology, Society, and Technology in the Citrus Industry of Palestine, 1890–1939*. Albany: State University of New York Press, 2005.

———. *Citrus Blossoms: Jewish Entrepreneurship in Palestine, 1890–1939* [Hebrew]. Jerusalem: Magnes, 2000.

Karmi, Dov. "Ha-Orientatzia Shel Ha-Dira Be-Tel Aviv." *Building in the Near East* 9–10 (November 1936): 5.

———. "Mibe'ayot Tikun Ha-Dirot Be-Tel Aviv." *Journal of the Association of Engineers and Architects* 7 (1945): 9–11.

Katz, Ethan, Lisa Moses Leff, and Maud Mandel. "Introduction: Engaging Colonial History and Jewish History." In *Colonialism and the Jews*, edited by Ethan Katz, Lisa Moses Leff, and Maud Mandel, 1–28. Bloomington: Indiana University Press, 2017.

Kazenelson, R. *Briut Ha-Ole: Madrikh La-Ole Ha-Chadash Be-Inyanei Higyena U-Vriut*. Jerusalem: Kupat Cholim Amamit Ve Machleket Ha-Aliya Shel Ha-Sokhnut Ha-Yehudit, 1939.

Kenan, Amos. *Your Land, Your Country* [Hebrew]. Tel Aviv: Yediot Aharonot, 1981.

Kennedy, Dane. "The Perils of the Midday Sun: Climatic Anxieties in the Colonial Tropics." In *Imperialism and the Natural World*, edited by John M. MacKenzie, 118–40. Manchester: Manchester University Press, 1990.

Khalidi, Rashid. *The Hundred Years' War on Palestine: A History of Settler Colonial Conquest and Resistance*. London: Profile Books, 2020.

Kligler, Israel Jacob. *The Epidemiology and Control of Malaria in Palestine*. Chicago: University of Chicago Press, 1930.

"Das Klima Palästinas." *Die Welt*, August 7, 1903.

Knox, Robert. *The Races of Men: A Fragment*. London: Henry Renshaw, 1850.

Kroyanker, David. *Jerusalem Architecture—Periods and Styles of Jewish Neighborhoods and Public Buildings Outside the City Walls* [Hebrew]. Tel Aviv: Keter, 1988.

———. *Jerusalem Architecture—Periods and Styles of the Period of the British Mandate 1918–1948* [Hebrew]. Tel Aviv: Keter, 1989.

Lafferty, Kevin D. "The Ecology of Climate Change and Infectious Diseases." *Ecology* 90, no. 4 (2009): 888–900.

Latour, Bruno. "How Better to Register the Agency of Things." The Tanner Lecture on Human Values 34 (2016 [March 26, 2014]): 79–117.

———. *We Have Never Been Modern*. Cambridge, MA: Harvard University Press, 2011.

Lehmann, Philipp. *Desert Edens: Colonial Climate Engineering in the Age of Anxiety*. Princeton: Princeton University Press, 2022.

Leimkugel, Frank. *Botanischer Zionismus: Otto Warburg (1859–1938) und die Anfänge institutionalisierter Naturwissenschaften in 'Erez Israel'*. Berlin:

veröffentlichungen aus dem botanischen Garten und botanischen Museum Berlin-Dahlem, 2005.

Levine, Mark. *Overthrowing Geography: Jaffa, Tel Aviv and the Struggle for Palestine 1880–1948*. Oakland: University of California Press, 2005.

Levy, M. *Merkaz Ha-Chizuyi Ve-Pe'ulotav* 16. Beit Dagan: Ha-Sherut Ha-Meteorologi, 1975.

Levy, Yael, and Nissim Levy. *The Physicians of the Holy Land, 1788–1948* [Hebrew]. Zikhron Ya'akov: Itay Bakhor, 2017.

Livingstone, N. David. "Race, Space and Moral Climatology: Notes Towards a Genealogy." *Journal of Historical Geography* 28, no. 2 (2002): 159–80.

Livneh, Eliezer. *Aaron Aaronsohn: Ha-Ish Ve-Zmano*. Jerusalem: Bialik, 1969.

Lowdermilk, Walter Clay. *Palestine Land of Promise*. London: Victor Gollancz, 1944.

Luck, J. "Climate Change and Diseases of Food Crops." *Plant Pathology* 60, no. 1 (2011): 113–21.

Mahony, Martin. "For an Empire of 'All Types of Climates': Meteorology as an Imperial Science." *Journal for Historical Geography* 51 (2016): 29–39.

Malki, Yossi. "Moderniyut, Leumiyut Ve-Hevra: Reshit Ha-Teufa Ha-Ivrit Be-Palestina- Eretz Israel." PhD diss., Tel Aviv University, 2007.

Mechkarim: Avodot Ha-Mosad Le-Cheker Ha-Bniya Ve-Hatekhnika. 2 vols. Jerusalem: The Association of Engineers, Architects and Measurers in Eretz Israel, 1944.

Meiton, Fredrik. *Electrical Palestine: Capital and Technology from Empire to Nation*. Oakland: University of California Press, 2019.

Melman, Billie. *Women's Orients: English Women and the Middle East, 1718–1918 : Sexuality, Religion and Work*. 2nd ed. Basingstoke, UK: Macmillan, 1995.

"Memazgei Avir Me-Israel Le-Rachavei Ha-Olam." *Ma'ariv*, July 12, 1956, 3.

Metzer, Jacob. *Economics, Land and Nationalism: Issues in Economic History and Political Economy in the Mandate Era and the State of Israel* [Hebrew]. Jerusalem: Magnes Press, 2023.

Metzger-Szmuk, Nitza. *Dwelling on the Dunes: Tel Aviv, Modern Movement and Bauhaus Ideals*. Paris: L'Éclat, 2004.

Meyer, Erna. *How to Cook in Palestine?* Tel Aviv: Palestine Federation of WIZO, 1937.

Mikhail, Alan. *Nature and Empire in Ottoman Egypt: An Environmental History*. Cambridge, UK: Cambridge University Press, 2011.

Misra, Anil Kumar. "Climate Change and Challenges of Water and Food Security." *International Journal of Sustainable Built Environment* 3, no. 1 (2014): 153–65.

Mitchell, Timothy. *Rule of Experts: Egypt, Techno-Politics, Modernity*. Berkeley: University of California Press, 2002.

"Mizug Avir Yutkan Be-Ein Gedi." *Al Ha-Mishmar*, July 22, 1956, 4.

Moon, David. "The Environmental History of the Russian Steppes: Vasilii Dockuchaev and the Harvest Failure of 1891." *Transactions of the Royal Historical Society* no. 15 (2005): 149–74.

Moulin, Anne Marie. "Tropical without the Tropics: The Turning-Point of Pastorian Medicine in North Africa." In *Warm Climates and Western Medicine: The Emergence of Tropical Medicine, 1500–1900*, edited by David Arnold, 160–80. Amsterdam: Rodopi, 1996.

Muthukumaran, Sureshkumar. *The Tropical Turn: Agricultural Innovation in the Ancient Middle East and the Mediterranean*. Berkeley: University of California Press, 2023.

Nadan, Amos. *The Palestinian Peasant Economy under the Mandate: A Story of Colonial Bungling*. Cambridge, MA: Harvard University Press, 2006.

Nash, Linda Lorraine. *Inescapable Ecologies: A History of Environment, Disease, and Knowledge*. Berkeley: University of California Press, 2006.

Nebeker, Frederik. *Calculating the Weather: Meteorology in the 20th Century*. San Diego: Academic Press, 1995.

Neill, Deborah Joy. *Networks in Tropical Medicine: Internationalism, Colonialism, and the Rise of a Medical Specialty, 1890–1930*. Stanford, CA: Stanford University Press, 2012.

Nitzan-Shiftan, Alona. "Disputes in Zionist Architecture Erich Mendelssohn and the Tel Aviv Hug." In *Architectural Culture: Place, Representation, Body* [Hebrew], edited by Rachel Kallus and Tali Hatuka, 201–29. Tel Aviv: Resling, 2005.

Norris, Jacob. *Land of Progress: Palestine in the Age of Colonial Development, 1905–1948*. Oxford: Oxford University Press, 2013.

Novick, Tamar. *Milk and Honey: Technologies of Plenty in the Making of a Holy Land*. Cambridge, MA: MIT Press, 2023.

Odlum, George M. "Gidul Ha-Bananot Be-Palestina." Department of Agriculture and Forestry, 1927.

Omissi, David E. *Air Power and Colonial Control: The Royal Air Force 1919–1939*. Manchester, UK: Manchester University Press, 1990.

Oppenheimer, Hanan. *Cultivation of Sub-Tropical and Tropical Fruit Trees* [Hebrew]. The Jewish Agency: Sifriyat Ha-Sade, 1942.

Osborne, Michael. *The Emergence of Tropical Medicine in France*. Chicago: University of Chicago Press, 2014.

Parolini, Giuditta. "Weather, Climate, and Agriculture: Historical Contributions and Perspectives from Agricultural Meteorology." *WIREs Climate Change* 13, no. 3 (2022): 1–19.

Pelli, Moshe. *Haskalah and Beyond: The Reception of the Hebrew Enlightenment and the Emergence of Haskalah Judaism*. Lanham, MD: University Press of America, 2010.

Penslar, Derek Jonathan. *Zionism and Technocracy: The Engineering of Jewish Settlement in Palestine, 1870–1918*. Bloomington: Indiana University Press, 1991.

———. "Zionism, Colonialism and Technocracy: Otto Warburg and the Commission for the Exploration of Palestine, 1903–7." *Journal of Contemporary History* 25, no. 1 (January 1990): 143–60.

Perry, Yaron, and Efraim Lev. *Modern Medicine in the Holy Land: Pioneering British Medical Services in Late Ottoman Palestine*. New York: Tauris Academic Studies, 2007.

Pinhas, Shira. "Road, Map: Partition in Palestine from the Local to the Transnational." *Journal of Levantine Studies* 10, no. 1 (2020): 111–21.

Piterberg, Gabriel. *The Returns of Zionism: Myths, Politics and Scholarship in Israel*. London: Verso, 2008.

Posener, Julius. "Batim Bnei Dira Achat Be-Eretz Israel." *Building* 1 (1934): 1–3.

Power for Palestine: The Rutenberg Concession. New York: Zionist Organization of America, 1922.

Qumsiyeh, Mazin B., and Mohammed A. Abusarhan. "An Environmental Nakba: The Palestinian Environment under Israeli Colonization." *Science Under Occupation* 23, no. 1 (Spring 2020): https://magazine.scienceforthepeople.org /vol23-1/an-environmental-nakba-the-palestinian-environment-under-israeli-colonization.

Ratzel, Friedrich. *Politische Geographie*. München und Leipzig: R. Oldenbourg 1897.

Raviv, Yael. *Falafel Nation: Cuisine and the Making of National Identity in Israel*. Lincoln: University of Nebraska Press, 2015.

Raz-Krakotzkin, Amnon. "En Elohim, Aval Hu Hevtiach Lanu Et Ha-Aretz." *Mita'am* 3 (2005): 71–76.

Rees, Jonathan. *Refrigeration Nation: A History of Ice, Appliances, and Enterprise in America*. Baltimore: John Hopkins University Press, 2013.

Robinson, Shira. *Citizen Strangers: Palestinians and the Birth of Israel's Liberal Settler State*. Stanford, CA: Stanford University Press, 2013.

Rohr, Jason R. "Frontiers in Climate Change–Disease Research." *Trends in Ecology & Evolution* 26, no. 6 (2011): 270–77.

Rosenberg, Elissa. "An All Day Garden: The Kibbutz as a Modernist Landscape." *Journal of Landscape Architecture* 7, no. 2 (2012): 32–39.

Rotbard, Sharon. *White City, Black City: Architecture and War in Tel Aviv and Jaffa*. London: Pluto Press, 2015.

Rovner, Adam. *In the Shadow of Zion: Promised Lands before Israel*. New York: NYU Press, 2014.

Ruppin, Arthur. "The Selection of the Fittest." In *Three Decades of Palestine: Speeches and Papers on the Upbuilding of the Jewish National Home*, 66–80. Jerusalem: Schocken, 1936.

Sabbagh-Khoury, Areej. "Tracing Settler Colonialism: A Genealogy of a Paradigm in the Sociology of Knowledge Production in Israel." *Politics and Society* 50, no. 1 (2022): 44–83.

Said, Edward W. *Orientalism*. New York: Penguin Books, 1995 [1978].

Samuel, Herbert. "An Interim Report on the Civil Administration of Palestine, During the Period 1st July, 1920–30th June, 1921." Edited by Office of Public Sector Information. London: HMSO, 1921.

———. *Report of the High Commissioner on the Administration of Palestine 1920–1925*. London: His Majesty's Stationery Office, 1925.

Sandler, Aaron. "The Health Conditions of Palestine." In *Zionist Work in Palestine*, edited by Israel Cohen, 73–85. New York: Judaean Publishing Company, 1912.

Schiebinger, Londa, and Claudia Swan, eds. *Colonial Botany: Science, Commerce, and Politics in the Early Modern World*. Philadelphia: University of Pennsylvania Press, 2005.

Schiller, Otto. "Ha-Avodot Ha-Mukdamot Letikhun Irony." *Building* 2 (1935): 1.

Scott, James C. *Seeing Like a State: How Certain Schemes to Improve the Human Condition Have Failed*. New Haven: Yale University Press, 1998.

Seide, Jacob. *Briut Ha-Am: Ha-Teva, Ha-Aklim, Ve-Ha-Briut Be-Israel*. Jerusalem: Reuven Mas, 1954.

Seikaly, Sherene. *Men of Capital: Scarcity and Economy in Mandate Palestine*. Stanford, CA: Stanford University Press, 2016.

Seligman, Noam G. "The Environmental Legacy of the Fallaheen and the Bedouin." In *Between Ruin and Restoration: An Environmental History of Israel*, edited by Daniel E. Orenstein, Char Miller and Alon Tal, 29–53. Pittsburgh: University of Pittsburgh Press, 2013.

Shadar, Hadas. *Sunstroke: Brutalist Construction in Be'er Sheva Re-Examination of National Architecture* [Hebrew]. Jerusalem: Yad Ben Zvi.

Shafir, Gershon. *Land, Labor and the Origins of the Israeli-Palestinian Conflict, 1882–1914*. Cambridge, UK: Cambridge University Press, 1989.

Shavit, Yaacov, and Mordechai Eran. *The Hebrew Bible Reborn: From Holy Scripture to the Book of Books; A History of Biblical Culture and the Battles over the Bible in Modern Judaism*. Berlin: Walter de Gruyter, 2007.

Shechter, M. "Al Ha-Ma'akhlim Ve-Ha-Mashkaot." *Davar*, May 5, 1936, 9.

———. "Aseret Ha-Dibrot Le-Yemei Ha-Hamsin." *Davar*, May 6, 1935.

Shilony, Zvi. *Jewish National Fund and Settlement in Eretz-Israel 1903–1914* [Hebrew]. Jerusalem: Yad Ben-Zvi, 1990.

Shoham, Hizky. *Carnival in Tel Aviv: Purim and the Celebration of Urban Zionism*. Boston, MA: Academic Studies Press, 2020.

Shorr, David. "Forest Law in Mandate Palestine." In *Managing the Unknown: Essays on Environmental Ignorance*, edited by Frank Uekotter and Uwe Luebken, 71–90. New York: Berghahn Books, 2014.

Smilansky, Moshe. "Tzel." *Ha-Aretz*, July 7, 1944, 3.

Smith, Andrea. "Queer Theory and Native Studies: The Heteronormativity of Settler Colonialism." *GLQ* 16, no. 1–2 (2010): 41–68.

Smith, Anthony D. *Chosen Peoples*. Oxford; New York: Oxford University Press, 2003.

Smith, Woodruff D. "Friedrich Ratzel and the Origins of Lebensraum." *German Studies Review* 3, no. 1 (1980): 51–68.

Soskin, S. E. *Small Holding and Irrigation: The New Form of Settlement in Palestine*. London: Jewish National Fund, 1920.

Stähler, Axel. *Zionism, the German Empire, and Africa: Jewish Metamorphoses and the Colors of Difference*. Oldenbourg: De Gruyter, 2019.

Stephan, Stephan Hanna. "The Division of the Year in Palestine." *The Journal of the Palestine Oriental Society* II (1922): 159–99.

Stoller, Shmuel. *Edut: Levatim Be-Hakamat Meshek Shlechin Be-Bika'at Kineret*. Tel Aviv: Hotza'at Ha-Merkaz Ha-Chaklai, 1980.

Strauss, Walter. "Ha-Aklim Ha-Cham Ve-Ha-Tilboshet." In *Ha-Adam Ve-Ha-Aklim Be-Eretz-Israel*, edited by S. Rosenbaum. Tel Aviv: Olympia, n.d.

Stübben, Josef. *Handbuch der Architektur: der Stadtbau*. Darmstadt: Arnold Bergstrasser, 1890.

Sufian, Sandra M. "Arab Health Care During the British Mandate, 1920–1947." In *Separate and Cooperate, Cooperate and Separate: The Disengagement of the Palestine Health Care System from Israel and Its Emergence as an Independent System*, edited by Tamara Barnea and Rafiq Husseini, 9–30. London: Praeger, 2002.

———. *Healing the Land and the Nation: Malaria and the Zionist Project in Palestine, 1920–1947*. Chicago: University of Chicago Press, 2007.

Tahon, Jakob. *Warburg's Book* [Hebrew]. Tel Aviv: Masada, 1948.

"Le-Takanat Ha-Batim Ha-Nivnim Be-Tel-Aviv." October 15, 1935. In *Tel Aviv—a Silver Jubilee: A Documentary Anthology* [Hebrew], edited by Maoz Azaryahu, Arnon Golan and Aminadav Dikman, 154–56, 75–81. Jerusalem: Carmel Publication, 2009.

Tal, Alon. *All the Trees of the Forest: Israel's Woodlands from the Bible to the Present*. New Haven: Yale University Press, 2013.

———. *Pollution in a Promised Land: An Environmental History of Israel*. Berkeley: University of California Press, 2002.

———. "To Make a Desert Bloom: The Israeli Agricultural Adventure and the Quest for Sustainability." *Agricultural History* 81, no. 2 (2007): 228–57.

Tamari, Salim. *Mountain against the Sea: Essays on Palestinian Society and Culture*. Berkeley: University of California Press, 2009.

Tene, Ofra. "The New Immigrant Must Not Only Learn." In *Jews and Their Foodways*, edited by Anat Helman, 46–64. Oxford: Oxford University Press, 2016.

———. "The White Houses Were Filled: Everyday Life in Homes in Tel Aviv 1924–1948" [Hebrew]. PhD diss., Tel Aviv University, 2010.

Tinovitch, T. "Hitpatchut Ha-Taasiya Shel Chomrei Binyan Ba-Aretz." *Journal of the Association of Engineers and Architects* 7 (1945): 3–8.

Triendl-Zadoff, Mirjam. *Next Year in Marienbad: The Lost Worlds of Jewish Spa Culture*. Philadelphia: University of Pennsylvania Press, 2012.

Tsur, Muky, and Yuval Danieli. *Mestechkin Builds Israel: Architecture in the Kibbutz* [Hebrew]. Tel Aviv: Ha-kibbutz Ha-Meuhad, 2008.

Tuchman, Barbara W. *Bible and Sword: England and Palestine from the Bronze Age to Balfour*. London: Papermac, 1982.

Valencius, Conevery Bolton. *The Health of the Country: How American Settlers Understood Themselves and Their Land*. New York: Basic Books, 2002.

Vaughan, Megan. *Curing Their Ills: Colonial Power and African Illness*. Cambridge, UK: Polity, 1991.

Veracini, Lorenzo. *Israel and Settler Society*. London: Pluto, 2006.

———. "The Other Shift: Settler Colonialism, Israel, and the Occupation." *Journal of Palestine Studies* 42, no. 2 (2013): 26–42.

———. *Settler Colonialism: A Theoretical Overview*. New York: Palgrave Macmillan, 2010.

von Suffrin, Dana. *Pflanzen für Palästina: Otto Warburg und die Naturwissenschaften im Jischuw*. Tübingen: Mohr Siebeck, 2019.

———. "The Possibility of a Productive Palestine: Otto Warburg and Botanical Zionism." *Israel Studies* 26, no. 2 (Summer 2021): 173–97.

Wachter, Clemens. *Die Professoren und Dozenten der Friedrich-Alexander-Universität Erlangen 1743–1960 (Teil 3)*. Erlangen: Universitätsbund Erlangen-Nürnberg, 2009.

Wahrman, Dror. *The Making of the Modern Self: Identity and Culture in Eighteenth-Century England*. New Haven: Yale University Press, 2004.

Warburg, Otto. "Die Bedeutung der Obstkulturen für die Kolonisation Palästinas." In *Tropische Landwirtschaft*. Berlin: Verlag von Gebrueder Borntraeger 1927/1907.

———. "Die nichtjüdische Kolonisation Palästinas." *Altneuland* 2 (February 1904): 39–45.

———. "Vegetation in Palestine." In *Zionist Work in Palestine*, edited by Israel Cohen. New York: Judaean Publication Company, 1912.

Weisbord, Robert G. *African Zion; the Attempt to Establish a Jewish Colony in the East Africa Protectorate, 1903–1905*. Philadelphia, Jewish Publication Society of America, 1968.

Weitz, Joseph. *Forest and Afforestation in Israel* [Hebrew]. Ramat Gan: Masada, 1970.

———. *Forest Policy in Israel* [Hebrew]. Jerusalem: Jewish National Fund Publication, 1950.

Weizman, Eyal. *The Conflict Shoreline: Colonization as Climate Change in the Negev Desert*. Göttingen: Steidl, 2015.

Wheeler, Roxann. *The Complexion of Race: Categories of Difference in Eighteenth-Century British Culture*. Philadelphia: University of Pennsylvania Press, 2000.

Whiteley, Nigel. *Reyner Banham: Historian of the Immediate Future*. Cambridge, MA: MIT Press, 2002.

Wilbush, Nahum. *Ha-Masa Le-Uganda*. Jerusalem: Ha-Sifriya Ha-Tzionit, 1963.

Wilson, Charles William. "On Recent Surveys in Sinai and Palestine." *Proceedings of the Royal Geographical Society of London* 17, no. 5 (1873): 326–34.

Wise, Norton M. *The Values of Precision*. Princeton: Princeton University Press, 1995.

Wittkower, Werner Joseph. "Aklim Be-Binyanei Ta'asiya Be-Eretz Israel." *Journal of the Association of Engineers and Architects* 7 (1945): 3.

Wolf, Aaron T. *Hydropolitics Along the Jordan River: Scarce Water and Its Impact on the Arab-Israeli Conflict*. Tokyo: United Nations University Press, 1995.

Wolfe, Patrick. "Land, Labor, and Difference: Elementary Structures of Race," *The American Historical Review* 106, no. 3 (2001): 866–905.

———. "Settler Colonialism and the Elimination of the Native." *Journal of Genocide Research* 8, no. 4 (2006): 387–409.

Woods, Rebecca J. H. "Nature and the Refrigerating Machine: The Politics and Production of Cold in the Nineteenth Century." In *Cryopolitics: Frozen Life in a Melting World*, edited by Joanna Radin and Emma Kowal, 89–116. Cambridge, MA: MIT Press, 2017.

Worboys, Michael. "The Colonial World as Mission and Mandate: Leprosy and Empire, 1900–1940." *Osiris* 15, no. 1 (2000): 207–18.

Wright, Glenn C. "Irrigating Citrus Trees." University of Arizona Cooperative Extension (February 2000): 1–5.

Yacobi, Haim. "Words and Place: The Case of the City Lod." In *Architectural Culture: Place, Representation, Body* [Hebrew], edited by Tali Hatuka and Rachel Kallus, 79–106. Tel Aviv: Resling, 2005.

Yaffe, Hilel. *Pirkei Zikhronot*. Tel Aviv: Unknown, 1935.

———. *Shmirat Ha'briut*. Jaffa: A. Atin Print, 1913.

Yaski, Yuval, and Galia Bar Or. *Kibbutz: Architecture without Precedents* [Hebrew]. Ein Harod: Museum of Art, 2010.

Yehudai, Ori. "Displaced in the National Home: Jewish Repatriation from Palestine to Europe, 1945–48." *Jewish Social Studies* 20, no. 2 (Winter 2014): 69–110.

———. *Leaving Zion: Jewish Emigration from Palestine and Israel after World War 2*. Cambridge, UK: Cambridge University Press, 2020.

Zahra, Tara. *The Great Departure: Mass Migration from Eastern Europe and the Making of the Free World*. New York: W. W. Norton and Company, 2016.

———. "Zionism, Emigration, and East European Colonialism." In *Colonialism and the Jews*, edited by Ethan Katz, Lisa Moses Leff and Maud Mandel, 166–92. Bloomington: Indiana University Press, 2017.

Zaidman, Miki, and Ruth Kark. "The Beginnings of Tel Aviv: The Achuzat Bayit Neighborhood as a 'Garden City'?" [Hebrew]. *Zmanim: A Historical Quarterly* 106 (Spring 2009): 8–19.

Zerubavel, Yael. *Desert in the Promised Land*. Stanford, CA: Stanford University Press, 2019.

Zeynep, Çelik. "Le Corbusier, Orientalism, Colonialism." *Assemblage: Critical Journal of Architecture and Design Culture* 17 (April 1992): 59–77.

———. *Urban Forms and Colonial Confrontations: Algiers under French Rule*. Berkeley: University of California Press, 1997.

Zikhronot Me-Eretz Israel. Edited by Yehoshua Bivar. Jerusalem: Ha-Machlaka Le-Chinukh U-Le-Tarbut Ba-Gola Shel Ha-Histadrut Ha-Tziyonit, 1985.

Zivan, Zeev. *From Nitsana to Eilat* [Hebrew]. Sde Boker: Ben-Gurion Research Institute for the Study of Israel and Zionism, 2012.

Zlocisti, Theodor. *Klimatologie und Pathologie Palästinas*. Tel Aviv: Omanut Co., 1937.

Zwarteveen, Margreet. "Transformations to Groundwater Sustainability: From Individuals and Pumps to Communities and Aquifers." *Current Opinion in Environmental Sustainability* 94 (April 2021): 88–97.

Index

Founded in 1893,
UNIVERSITY OF CALIFORNIA PRESS
publishes bold, progressive books and journals
on topics in the arts, humanities, social sciences,
and natural sciences—with a focus on social
justice issues—that inspire thought and action
among readers worldwide.

The UC PRESS FOUNDATION
raises funds to uphold the press's vital role
as an independent, nonprofit publisher, and
receives philanthropic support from a wide
range of individuals and institutions—and from
committed readers like you. To learn more, visit
ucpress.edu/supportus.